NYSTCE
004

CST
Mathematics
Teacher Certification Exam

By: Sharon Wynne, M.S
Southern Connecticut State University

"And, while there's no reason yet to panic, I think it's only prudent that we make preparations to panic."

XAMonline, INC.
Boston

To obtain permission(s) to use the material from this work for any purpose including workshops or seminars, please submit a written request to:

XAMonline, Inc.
21 Orient Ave.
Melrose, MA 02176
Toll Free 1-800-509-4128
Email: info@xamonline.com
Web www.xamonline.com
Fax: 1-781-662-9268

Library of Congress Cataloging-in-Publication Data

Wynne, Sharon A.
    CST Mathematics 004: Teacher Certification / Sharon A. Wynne. -2nd ed.
    ISBN 978-1-58197-296-2
    1. CST Mathematics 004.   2. Study Guides.   3. NYSTCE
    4. Teachers' Certification & Licensure.   5. Careers

Disclaimer:

The opinions expressed in this publication are the sole works of XAMonline and were created independently from the National Education Association, Educational Testing Service, or any State Department of Education, National Evaluation Systems or other testing affiliates.

Between the time of publication and printing, state specific standards as well as testing formats and website information may change that is not included in part or in whole within this product. Sample test questions are developed by XAMonline and reflect similar content as on real tests; however, they are not former tests. XAMonline assembles content that aligns with state standards but makes no claims nor guarantees teacher candidates a passing score. Numerical scores are determined by testing companies such as NES or ETS and then are compared with individual state standards. A passing score varies from state to state.

Printed in the United States of America

NYSTCE: CST Mathematics 004
ISBN: 978-1-58197-296-2

About the Subject Assessments

NYSTCE™: Subject Assessment in the Mathematics examination

Purpose: The assessments are designed to test the knowledge and competencies of prospective secondary level teachers. The question bank from which the assessment is drawn is undergoing constant revision. As a result, your test may include questions that will not count towards your score.

Test Version: There is one version of subject assessment for Mathematics in New York. The Mathematics (004) exam emphasizes comprehension in Mathematic Reasoning and Communication; Algebra; Trigonometry and Calculus; Measurement and Geometry; Data Analysis, Statistics and Discrete Mathematics. The Mathematics study guide is based on a typical knowledge level of persons who have completed a *bachelor's degree program* in Mathematics.

Time Allowance: You will have 4 hours to finish the exam. There exam consists of approximately 90 multiple-choice questions in the exam and 1 constructed-response assignment.

Weighting: Approximately 13% of the test material consists of multiple-choice questions in Mathematic Reasoning and Communication; 26% consists of multiple-choice questions in Algebra; 17% consists of multiple-choice questions in Trigonometry and Calculus; 17% consists of multiple-choice questions in Measurement and Geometry; 17% consists of multiple-choice questions in Data Analysis, Statistics and Discrete Mathematics; 10% consists of a constructed-response assignment in Algebra.

Additional Information about the NYSTCE Assessments: The NYSTCE series subject assessments are developed by *National Evaluation Systems.* They provide additional information on the NYSTCE series assessments, including registration, preparation and testing procedures and study materials such topical guides that have about 51 pages of information including approximately 22 additional sample questions.

TABLE OF CONTENTS

PG.

Great Study and Testing Tips!

What to study in order to prepare for the subject assessments is the focus of this study guide but equally important is *how* you study.

You can increase your chances of truly mastering the information by taking some simple, but effective steps.

Study Tips:

1. **Some foods aid the learning process.** Foods such as milk, nuts, seeds, rice, and oats help your study efforts by releasing natural memory enhancers called CCKs (*cholecystokinin*) composed of *tryptopha*n, *choline*, and *phenylalanine*. All of these chemicals enhance the neurotransmitters associated with memory. Before studying, try a light, protein-rich meal of eggs, turkey, and fish. All of these foods release the memory enhancing chemicals. The better the connections, the more you comprehend.

Likewise, before you take a test, stick to a light snack of energy boosting and relaxing foods. A glass of milk, a piece of fruit, or some peanuts all release various memory-boosting chemicals and help you to relax and focus on the subject at hand.

2. **Learn to take great notes.** A by-product of our modern culture is that we have grown accustomed to getting our information in short doses (i.e. TV news sound bites or USA Today style newspaper articles.)

Consequently, we've subconsciously trained ourselves to assimilate information better in neat little packages. If your notes are scrawled all over the paper, it fragments the flow of the information. Strive for clarity. Newspapers use a standard format to achieve clarity. Your notes can be much clearer through use of proper formatting. A very effective format is called the *"Cornell Method."*

Take a sheet of loose-leaf lined notebook paper and draw a line all the way down the paper about 1-2" from the left-hand edge.

Draw another line across the width of the paper about 1-2" up from the bottom. Repeat this process on the reverse side of the page.

Look at the highly effective result. You have ample room for notes, a left hand margin for special emphasis items or inserting supplementary data from the textbook, a large area at the bottom for a brief summary, and a little rectangular space for just about anything you want.

3. <u>Get the concept then the details</u>. Too often we focus on the details and don't gather an understanding of the concept. However, if you simply memorize only dates, places, or names, you may well miss the whole point of the subject.

A key way to understand things is to put them in your own words. If you are working from a textbook, automatically summarize each paragraph in your mind. If you are outlining text, don't simply copy the author's words.

Rephrase them in your own words. You remember your own thoughts and words much better than someone else's, and subconsciously tend to associate the important details to the core concepts.

4. <u>Ask Why?</u> Pull apart written material paragraph by paragraph and don't forget the captions under the illustrations.

Example: If the heading is "Stream Erosion", flip it around to read "Why do streams erode?" Then answer the questions.

If you train your mind to think in a series of questions and answers, not only will you learn more, but it also helps to lessen the test anxiety because you are used to answering questions.

5. <u>Read for reinforcement and future needs</u>. Even if you only have 10 minutes, put your notes or a book in your hand. Your mind is similar to a computer; you have to input data in order to have it processed. *By reading, you are creating the neural connections for future retrieval.* The more times you read something, the more you reinforce the learning of ideas.

Even if you don't fully understand something on the first pass, *your mind stores much of the material for later recall.*

6. <u>Relax to learn so go into exile</u>. Our bodies respond to an inner clock called biorhythms. Burning the midnight oil works well for some people, but not everyone.

If possible, set aside a particular place to study that is free of distractions. Shut off the television, cell phone, pager and exile your friends and family during your study period.

If you really are bothered by silence, try background music. Light classical music at a low volume has been shown to aid in concentration over other types. Music that evokes pleasant emotions without lyrics are highly suggested. Try just about anything by Mozart. It relaxes you.

7. <u>Use arrows not highlighters</u>. At best, it's difficult to read a page full of yellow, pink, blue, and green streaks. Try staring at a neon sign for a while and you'll soon see that the horde of colors obscure the message.

A quick note, a brief dash of color, an underline, and an arrow pointing to a particular passage is much clearer than a horde of highlighted words.

8. <u>Budget your study time</u>. Although you shouldn't ignore any of the material, *allocate your available study time in the same ratio that topics may appear on the test.*

Testing Tips:

1. <u>**Get smart, play dumb.**</u> **Don't read anything into the question.** Don't make an assumption that the test writer is looking for something else than what is asked. Stick to the question as written and don't read extra things into it.

2. <u>**Read the question and all the choices *twice* before answering the question.**</u> You may miss something by not carefully reading, and then re-reading both the question and the answers.

If you really don't have a clue as to the right answer, leave it blank on the first time through. Go on to the other questions, as they may provide a clue as to how to answer the skipped questions.

If later on, you still can't answer the skipped ones . . . ***Guess.*** The only penalty for guessing is that you *might* get it wrong. Only one thing is certain; if you don't put anything down, you will get it wrong!

3. <u>**Turn the question into a statement.**</u> Look at the way the questions are worded. The syntax of the question usually provides a clue. Does it seem more familiar as a statement rather than as a question? Does it sound strange?

By turning a question into a statement, you may be able to spot if an answer sounds right, and it may also trigger memories of material you have read.

4. <u>**Look for hidden clues.**</u> It's actually very difficult to compose multiple-foil (choice) questions without giving away part of the answer in the options presented.

In most multiple-choice questions you can often readily eliminate one or two of the potential answers. This leaves you with only two real possibilities and automatically your odds go to Fifty-Fifty for very little work.

5. <u>**Trust your instincts.**</u> For every fact that you have read, you subconsciously retain something of that knowledge. On questions that you aren't really certain about, go with your basic instincts. **Your first impression on how to answer a question is usually correct.**

6. <u>**Mark your answers directly on the test booklet**</u>. Don't bother trying to fill in the optical scan sheet on the first pass through the test.

Just be very careful not to miss-mark your answers when you eventually transcribe them to the scan sheet.

7. <u>**Watch the clock**</u>! You have a set amount of time to answer the questions. Don't get bogged down trying to answer a single question at the expense of 10 questions you can more readily answer.

SUBAREA I–MATHEMATIC REASONING AND COMMUNICATION

0001. Understand reasoning processes, including inductive and deductive logic and symbolic logic.

Conditional statements are frequently written in "**if-then**" form. The "if" clause of the conditional is known as the **hypothesis**, and the "then" clause is called the **conclusion**. In a proof, the hypothesis is the information that is assumed to be true, while the conclusion is what is to be proven true. A conditional is considered to be of the form:

If p, then q
p is the hypothesis. q is the conclusion.

Conditional statements can be diagrammed using a **Venn diagram**. A diagram can be drawn with one circle inside another circle. The inner circle represents the hypothesis. The outer circle represents the conclusion. If the hypothesis is taken to be true, then you are located inside the inner circle. If you are located in the inner circle then you are also inside the outer circle, so that proves the conclusion is true.

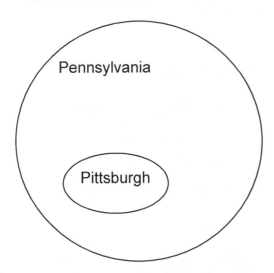

Example:
If an angle has a measure of 90 degrees, then it is a right angle.

In this statement "an angle has a measure of 90 degrees" is the hypothesis.
In this statement "it is a right angle" is the conclusion.

Example:
If you are in Pittsburgh, then you are in Pennsylvania.
In this statement "you are in Pittsburgh" is the hypothesis.
In this statement "you are in Pennsylvania" is the conclusion.

Conditional: If p, then q	p is the hypothesis. q is the conclusion.
Inverse: If ~ p, then ~ q.	Negate both the hypothesis (If not p, then not q) and the conclusion from the original conditional.
Converse : If q, then p.	Reverse the 2 clauses. The original hypothesis becomes the conclusion. The original conclusion then becomes the new hypothesis.
Contrapositive: If ~ q, then ~ p.	Reverse the 2 clauses. The If not q, then not p original hypothesis becomes the conclusion. The original conclusion then becomes the new hypothesis. THEN negate both the new hypothesis and the new conclusion.

<u>Example</u>: Given the **conditional**:

If an angle has 60°, then it is an acute angle.

Its **inverse**, in the form "If ~ p, then ~ q", would be:

If an angle doesn't have 60°, then it is not an acute angle.

NOTICE that the inverse is not true, even though the conditional statement was true.

Its **converse**, in the form "If q, then p", would be:

If an angle is an acute angle, then it has 60°.

NOTICE that the converse is not true, even though the conditional statement was true.

Its **contrapositive**, in the form "If ~q, then ~p", would be:

If an angle isn't an acute angle, then it doesn't have 60°.

NOTICE that the contrapositive is true, assuming the original conditional statement was true.

TIP: If you are asked to pick a statement that is logically equivalent to a given conditional, look for the contra-positive. The inverse and converse are not always logically equivalent to every conditional. The contra-positive is ALWAYS logically equivalent.

Find the inverse, converse and contrapositive of the following conditional statement. Also determine if each of the 4 statements is true or false.

Conditional: If $x = 5$, then $x^2 - 25 = 0$.   TRUE
Inverse: If $x \neq 5$, then $x^2 - 25 \neq 0$.   FALSE, x could be ⁻5
Converse: If $x^2 - 25 = 0$, then $x = 5$.   FALSE, x could be ⁻5
Contrapositive: If $x^2 - 25 \neq 0$, then $x \neq 5$. TRUE

Conditional: If $x = 5$, then $6x = 30$.   TRUE
Inverse: If $x \neq 5$, then $6x \neq 30$.   TRUE
Converse: If $6x = 30$, then $x = 5$.   TRUE
Contrapositive: If $6x \neq 30$, then $x \neq 5$. TRUE

Sometimes, as in this example, all 4 statements can be logically equivalent; however, the only statement that will always be logically equivalent to the original conditional is the contrapositive.

Conditional statements can be diagrammed using a **Venn diagram**. A diagram can be drawn with one figure inside another figure. The inner figure represents the hypothesis. The outer figure represents the conclusion. If the hypothesis is taken to be true, then you are located inside the inner figure. If you are located in the inner figure then you are also inside the outer figure, so that proves the conclusion is true. Sometimes that conclusion can then be used as the hypothesis for another conditional, which can result in a second conclusion.

Suppose that these statements were given to you, and you are asked to try to reach a conclusion. The statements are:

All swimmers are athletes.
All athletes are scholars.

In "if-then" form, these would be:
   If you are a swimmer, then you are an athlete.
   If you are an athlete, then you are a scholar.

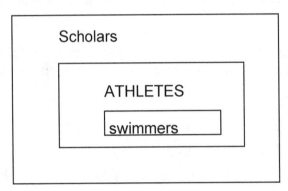

Clearly, if you are a swimmer, then you are also an athlete. This includes you in the group of scholars.

Suppose that these statements were given to you, and you are asked to try to reach a conclusion. The statements are:

All swimmers are athletes.
All wrestlers are athletes.

In "if-then" form, these would be:
   If you are a swimmer, then you are an athlete.
   If you are a wrestler, then you are an athlete.

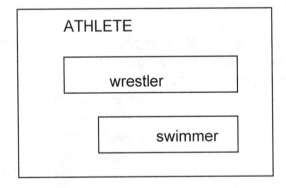

Clearly, if you are a swimmer or a wrestler, then you are also an athlete. This does NOT allow you to come to any other conclusions.

A swimmer may or may NOT also be a wrestler. Therefore, NO CONCLUSION IS POSSIBLE.

Suppose that these statements were given to you, and you are asked to try to reach a conclusion. The statements are:

All rectangles are parallelograms.
Quadrilateral ABCD is not a parallelogram.

In "if-then" form, the first statement would be:
If a figure is a rectangle, then it is also a parallelogram.

Note that the second statement is the negation of the conclusion of statement one. Remember also that the contrapositive is logically equivalent to a given conditional. That is, **"If ~ q, then ~ p"**. Since" ABCD is NOT a parallelogram " is like saying **"If ~ q,"** then you can come to the conclusion **"then ~ p"**. Therefore, the conclusion is ABCD is not a rectangle. Looking at the Venn diagram below, if all rectangles are parallelograms, then rectangles are included as part of the parallelograms. Since quadrilateral ABCD is not a parallelogram, that it is excluded from anywhere inside the parallelogram box. This allows you to conclude that ABCD can not be a rectangle either.

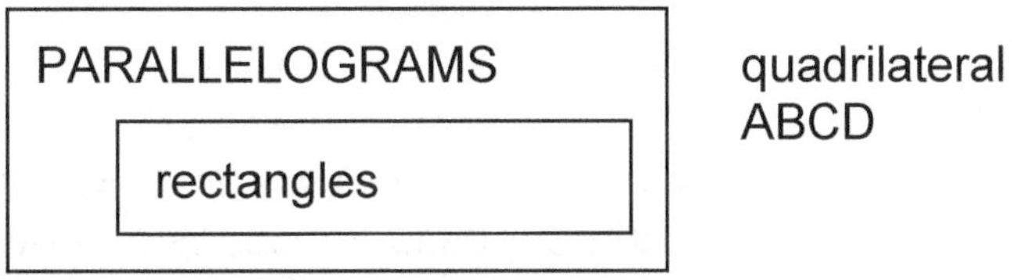

PARALLELOGRAMS

rectangles

quadrilateral
ABCD

Try These:

What conclusion, if any, can be reached? Assume each statement is true, regardless of any personal beliefs.

1. If the Red Sox win the World Series, I will die.
   I died.

2. If an angle's measure is between 0° and 90°, then the angle is acute. Angle B is not acute.

3. Students who do well in geometry will succeed in college.
   Annie is doing extremely well in geometry.

4. Left-handed people are witty and charming.
   You are left-handed.

The only **undefined terms** are point, line and plane.

Definitions are explanations of all mathematical terms except those that are undefined.

Postulates are mathematical statements that are accepted as true statements without providing a proof.

Theorems are mathematical statements that can be proven to be true based on postulates, definitions, algebraic properties, given information, and previously proved theorems.

The **3 undefined terms of geometry** are point, line, and plane.

A plane is a flat surface that extends forever in two dimensions. It has no ends or edges. It has no thickness to it. It is usually drawn as a parallelogram that can be named either by 3 non-collinear points (3 points that are not on the same line) on the plane or by placing a letter in the corner of the plane that is not used elsewhere in the diagram.

A line extends forever in one dimension. It is determined and named by 2 points that are on the line. The line consists of every point that is between those 2 points as well as the points that are on the "straight" extension each way. A line is drawn as a line segment with arrows facing opposite directions on each end to indicate that the line continues in both directions forever.

A point is a position in space, on a line, or on a plane. It has no thickness and no width. Only 1 line can go through any 2 points. A point is represented by a dot named by a single letter.

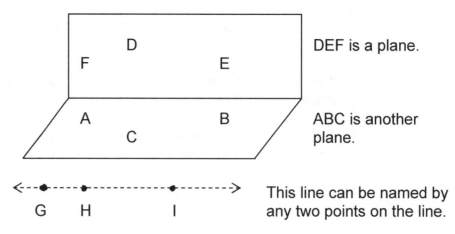

DEF is a plane.

ABC is another plane.

This line can be named by any two points on the line.

The line could be named $\overleftrightarrow{GH}$, $\overleftrightarrow{HI}$, $\overleftrightarrow{GI}$, $\overleftrightarrow{IG}$, $\overleftrightarrow{IH}$, or $\overleftrightarrow{HG}$. Any 2 points (letters) on the line can be used and their order is not important in naming a line.

In the previous diagrams, A, B, C, D, E, F, G, H, and I are all locations of individual points.

A ray is not an undefined term. A ray consists of all the points on a line starting at one given point and extending in only one of the two opposite directions along the line. The ray is named by naming 2 points on the ray. The first point must be the endpoint of the ray, while the second point can be any other point along the ray. The symbol for a ray is a ray above the 2 letters used to name it. The endpoint of the ray MUST be the first letter.

This ray could be named $\overrightarrow{JK}$ or $\overrightarrow{JL}$. It can not be called $\overrightarrow{KJ}$ or $\overrightarrow{LJ}$ or $\overrightarrow{LK}$ or $\overrightarrow{KL}$ because none of those names start with the endpoint, J.

The **distance** between 2 points on a number line is equal to the absolute value of the difference of the two numbers associated with the points.

If one point is located at "a" and the other point is at "b", then the distance between them is found by this formula:

$$\text{distance} = |a - b| \text{ or } |b - a|$$

If one point is located at $^-3$ and another point is located at 5, the distance between them is found by:

$$\text{distance} = |a - b| = |(^-3) - 5| = |^-8| = 8$$

In a **2 column proof**, the left side of the proof should be the given information, or statements that could be proved by deductive reasoning. The right column of the proof consists of the reasons used to determine that each statement to the left was verifiably true. The right side can identify given information, or state theorems, postulates, definitions or algebraic properties used to prove that particular line of the proof is true.

Assume the opposite of the conclusion. Keep your hypothesis given information the same. Proceed to develop the steps of the proof, looking for a statement that contradicts your original assumption or some other known fact. This contradiction indicates that the assumption you made at the beginning of the proof was incorrect; therefore, the original conclusion has to be true.

The following **algebraic postulates** are frequently used as reasons for statements in 2 column geometric properties:

Addition Property:

If $a = b$ and $c = d$, then $a + c = b + d$.

Subtraction Property:

If $a = b$ and $c = d$, then $a - c = b - d$.

Multiplication Property:

If $a = b$ and $c \neq 0$, then $ac = bc$.

Division Property:

If $a = b$ and $c \neq 0$, then $a/c = b/c$.

Reflexive Property: $a = a$
Symmetric Property: If $a = b$, then $b = a$.
Transitive Property: If $a = b$ and $b = c$, then $a = c$.
Distributive Property: $a(b + c) = ab + ac$
Substitution Property: If $a = b$, then b may be substituted for a in any other expression (a may also be substituted for b).

Inductive thinking is the process of finding a pattern from a group of examples. That pattern is the conclusion that this set of examples seemed to indicate. It may be a correct conclusion or it may be an incorrect conclusion because other examples may not follow the predicted pattern.

Deductive thinking is the process of arriving at a conclusion based on other statements that are all known to be true, such as theorems, axiomspostulates, or postulates. Conclusions found by deductive thinking based on true statements will **always** be true.

Examples :

Suppose:
   On Monday Mr.Peterson eats breakfast at McDonalds.
   On Tuesday Mr.Peterson eats breakfast at McDonalds.
   On Wednesday Mr.Peterson eats breakfast at McDonalds.
   On Thursday Mr.Peterson eats breakfast at McDonalds again.

Conclusion: On Friday Mr. Peterson will eat breakfast at McDonalds again.

This is a conclusion based on inductive reasoning. Based on several days observations, you conclude that Mr. Peterson will eat at McDonalds. This may or may not be true, but it is a conclusion arrived at by inductive thinking.

0002. Understand the meaning of mathematical concepts and symbols and how to communicate mathematical ideas in writing.

The unit rate for purchasing an item is its price divided by the number of pounds/ ounces, etc. in the item. The item with the lower unit rate is the lower price.

<u>Example:</u> Find the item with the best unit price:

$1.79 for 10 ounces
$1.89 for 12 ounces
$5.49 for 32 ounces

$$\frac{1.79}{10} = .179 \text{ per ounce} \qquad \frac{1.89}{12} = .1575 \text{ per ounce} \qquad \frac{5.49}{32} = .172 \text{ per ounce}$$

$1.89 for 12 ounces is the best price.

A second way to find the better buy is to make a proportion with the price over the number of ounces, etc. Cross multiply the proportion, writing the products above the numerator that is used. The better price will have the smaller product.

<u>Example:</u> Find the better buy:

$8.19 for 40 pounds or $4.89 for 22 pounds

Find the unit price.

$$\frac{40}{8.19} = \frac{1}{x} \qquad\qquad \frac{22}{4.89} = \frac{1}{x}$$
$$40x = 8.19 \qquad\qquad 22x = 4.89$$
$$x = .20475 \qquad\qquad x = .22\overline{227}$$

Since $.20475 < .22\overline{227}$, $8.19 is less and is a better buy.

To find the amount of sales tax on an item, change the percent of sales tax into an equivalent decimal number. Then multiply the decimal number times the price of the object to find the sales tax. The total cost of an item will be the price of the item plus the sales tax.

Example: A guitar costs $120 plus 7% sales tax. How much are the sales tax and the total bill?

7% = .07 as a decimal   (.07)(120) = $8.40 sales tax
$120 + $8.40 = $128.40 ← total price

Example: A suit costs $450 plus 6½% sales tax. How much are the sales tax and the total bill?

6½% = .065 as a decimal
(.065)(450) = $29.25 sales tax
$450 + $29.25 = $479.25 ← total price

Elapsed time problems are usually one of two types. One type of problem is the elapsed time between 2 times given in hours, minutes, and seconds. The other common type of problem is between 2 times given in months and years.

For any time of day past noon, change it into military time by adding 12 hours. For instance, 1:15 p.m. would be 13:15. Remember when you borrow a minute or an hour in a subtraction problem that you have borrowed 60 more seconds or minutes.

Example: Find the time from 11:34:22 a.m. until 3:28:40 p.m.

First change 3:28:40 p.m. to 15:28:40 p.m.
Now subtract   - 11:34:22 a.m.
                                        :18

Borrow an hour and add 60 more minutes. Subtract
14:88:40 p.m.
- 11:34:22 a.m.
3:54:18 ↔ 3 hours, 54 minutes, 18 seconds

<u>Example:</u> John lived in Arizona from September 91 until March 95. How long is that?

		year	month
March 95	=	95	03
September 91	= -	91	09

Borrow a year, change it into 12 more months, and subtract.

		year	month
March 95	=	94	15
September 91	= -	91	09
		3 yr	6 months

<u>Example:</u> A race took the winner 1 hr. 58 min. 12 sec. on the first half of the race and 2 hr. 9 min. 57 sec. on the second half of the race. How much time did the entire race take?

    1 hr. 58 min. 12 sec.
  + 2 hr. 9 min. 57 sec.   Add these numbers
    3 hr. 67 min. 69 sec.
  + 1 min -60 sec.   Change 60 seconds to 1 min.

    3 hr. 68 min. 9 sec.
  + 1 hr.-60 min.   .   Change 60 minutes to 1 hr.
    4 hr. 8 min. 9 sec.   ←final answer

To **convert a fraction to a decimal,** simply divide the numerator (top) by the denominator (bottom). Use long division if necessary.

If a decimal has a fixed number of digits, the decimal is said to be terminating. To write such a decimal as a fraction, first determine what place value the farthest right digit is in, for example: tenths, hundredths, thousandths, ten thousandths, hundred thousands, etc. Then drop the decimal and place the string of digits over the number given by the place value.

If a decimal continues forever by repeating a string of digits, the decimal is said to be repeating. To write a repeating decimal as a fraction, follow these steps.

a.   Let x = the repeating decimal
     (ex. $x = .716716716...$)
b.   Multiply x by the multiple of ten that will move the decimal just to the right of the repeating block of digits.
     (ex. $1000x = 716.716716...$)

c.   Subtract the first equation from the second.
      (ex. $1000x - x = 716.716.716... - .716716...$)

d.   Simplify and solve this equation. The repeating block of
      digits will subtract out.
      (ex. $999x = 716$ so $x = 716/999$)

e.   The solution will be the fraction for the repeating decimal.

To change a number into scientific notation, move the decimal point
so that only one number from 1 to 9 is in front of the decimal point.
Drop off any trailing zeros. Multiply this number times 10 to a
power. The power is the number of positions that the decimal point
is moved. The power is negative if the original number is a decimal
number between 1 and -1. Otherwise the power is positive.

<u>Example:</u> Change into scientific notation:

4,380,000,000	Move decimal behind the 4
4.38	Drop trailing zeros.
$4.38 \times 10^?$	Count positions that the decimal point has moved.
4.38×10^9	This is the answer.
$^-.0000407$	Move decimal behind the 4
$^-4.07$	Count positions that the decimal point has moved.
$^-4.07 \times 10^{-5}$	Note negative exponent.

A **ratio** is a comparison of 2 numbers. If a class had 11 boys and
14 girls, the ratio of boys to girls could be written one of 3 ways:

$$11:14 \quad \text{or} \quad 11 \text{ to } 14 \quad \text{or} \quad \frac{11}{14}$$

The ratio of girls to boys is:

$$14:11, \ 14 \text{ to } 11 \text{ or} \quad \frac{14}{11}$$

Ratios can be reduced when possible. A ratio of 12 cats to 18 dogs
would reduce to 2:3, 2 to 3 or $2/3$.

Note: Read ratio questions carefully. Given a group of 6 adults and
5 children, the ratio of children to the entire group would be 5:11.

A **proportion** is an equation in which a fraction is set equal to another. To solve the proportion, multiply each numerator times the other fraction's denominator. Set these two products equal to each other and solve the resulting equation. This is called **cross-multiplying** the proportion.

Example: $\dfrac{4}{15} = \dfrac{x}{60}$ is a proportion.

To solve this, cross multiply.

$$(4)(60) = (15)(x)$$
$$240 = 15x$$
$$16 = x$$

Example: $\dfrac{x+3}{3x+4} = \dfrac{2}{5}$ is a proportion.

To solve, cross multiply.

$$5(x+3) = 2(3x+4)$$
$$5x+15 = 6x+8$$
$$7 = x$$

Example: $\dfrac{x+2}{8} = \dfrac{2}{x-4}$ is another proportion.

To solve, cross multiply.

$$(x+2)(x-4) = 8(2)$$
$$x^2 - 2x - 8 = 16$$
$$x^2 - 2x - 24 = 0$$
$$(x-6)(x+4) = 0$$
$$x = 6 \text{ or } x = {}^-4$$

Both answers work.

If a number is already in scientific notation, it can be changed back into the regular decimal form. If the exponent on the number 10 is negative, move the decimal point to the left. If the exponent on the number 10 is positive, move the decimal point to the right that number of places.

<u>Example:</u> Change back into decimal form:

3.448×10^{-2}	Move decimal point 2 places left, since exponent is negative.
.03448	This is the answer.
6×10^4	Move decimal point 4 places right, since exponent is negative.
60,000	This is the answer.

To add or subtract in scientific notation, the exponents must be the same. Then add the decimal portions, keeping the power of 10 the same. Then move the decimal point and adjust the exponent to keep the number in front of the decimal point from 1 to 9.

<u>Example:</u>

6.22×10^3	
$+ \ 7.48 \times 10^3$	Add these as is.
13.70×10^3	Now move decimal 1 more place to the left and
1.37×10^4	add 1 more exponent.

To multiply or divide in scientific notation, multiply or divide the decimal part of the numbers. In multiplication, add the exponents of 10. In division, subtract the exponents of 10. Then move the decimal point and adjust the exponent to keep the number in front of the decimal point from 1 to 9.

Example:

$(5.2 \times 10^5)(3.5 \times 10^2)$ Multiply $5.2 \cdot 3.5$

18.2×10^7 Add exponent

1.82×10^8 Move decimal point and increase the exponent by 1.

Example:

$$\frac{(4.1076 \times 10^3)}{2.8 \times 10^{-4}}$$ Divide 4.1076 by 2.8

Subtract $3 - (^-4)$

1.467×10^7

The **real number properties** are best explained in terms of a small set of numbers. For each property, a given set will be provided.

Axioms of Addition

Closure—For all real numbers a and b, $a + b$ is a unique real number.

Associative—For all real numbers a, b, and c, $(a + b) + c = a + (b + c)$.

Additive Identity—There exists a unique real number 0 (zero) such that $a + 0 = 0 + a = a$ for every real number a.

Additive Inverses—For each real number a, there exists a real number $–a$ (the opposite of a) such that $a + (-a) = (-a) + a = 0$.

Commutative—For all real numbers a and b, $a + b = b + a$.

Axioms of Multiplication

Closure—For all real numbers a and b, ab is a unique real number.

Associative—For all real numbers a, b, and c, $(ab)c = a(bc)$.

Multiplicative Identity—There exists a unique nonzero real number 1 (one) such that $1 \cdot a = a \cdot 1$.

Multiplicative Inverses—For each nonzero real number, there exists a real number $1/a$ (the reciprocal of a) such that $a(1/a) = (1/a)a = 1$.

Commutative—For all real numbers a and b, $ab = ba$.

The Distributive Axiom of Multiplication over Addition

For all real numbers a, b, and c, $a(b + c) = ab + ac$.
Most numbers in mathematics are "exact" or "counted". Measurements are "approximate". They usually involve interpolation or figuring out which mark on the ruler is closest. Any measurement you get with a measuring device is approximate. Variations in measurement are called precision and accuracy.

Precision is a measurement of how exactly a measurement is made, without reference to a true or real value. If a measurement is precise it can be made again and again with little variation in the result. The precision of a measuring device is the smallest fractional or decimal division on the instrument. The smaller the unit or fraction of a unit on the measuring device, the more precisely it can measure.

The greatest possible error of measurement is always equal to one-half the smallest fraction of a unit on the measuring device.

Accuracy is a measure of how close the result of measurement comes to the "true" value.

If you are throwing darts, the true value is the bull's eye. If the three darts land on the bull's eye, the dart thrower is both precise (all land near the same spot) and accurate (the darts all land on the "true" value). The greatest measure of error allowed is called the tolerance. The least acceptable limit is called the lower limit and the greatest acceptable limit is called the upper limit. The difference between the upper and lower limits is called the tolerance interval. For example, a specification for an automobile part might be 14.625 ± 0.005 mm. This means that the smallest acceptable length of the part is 14.620 mm and the largest length acceptable is 14.630 mm. The tolerance interval is 0.010 mm. One can see how it would be important for automobile parts to be within a set of limits in terms of length. If the part is too long or too short it will not fit properly and vibrations will occur weakening the part and eventually causing damage to other parts.

0003. Understand mathematical modeling and apply multiple mathematical representations to connect mathematical ideas and solve problems.

Estimation and approximation may be used to check the reasonableness of answers.

<u>Example</u>: Estimate the answer.

$$\frac{58 \times 810}{1989}$$

58 becomes 60, 810 becomes 800 and 1989 becomes 2000.

$$\frac{60 \times 800}{2000} = 24$$

Word problems: An estimate may sometimes be all that is needed to solve a problem.

<u>Example</u>: Janet goes into a store to purchase a CD on sale for $13.95. While shopping, she sees two pairs of shoes, prices $19.95 and $14.50. She only has $50. Can she purchase everything?

Solve by rounding:

$19.95 → $20.00
$14.50 → $15.00
$13.95 → $14.00
$49.00   Yes, she can purchase the CD and the shoes.

Geometric Problems

The strategy for solving problems of this nature should be to identify the given shapes and choose the correct formulas. Subtract the smaller cut out shape from the larger shape.

Sample problems:

1. Find the area of one side of the metal in the circular flat washer shown below:

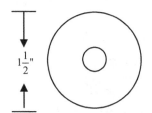

1. the shapes are both circles.

2. use the formula $A = \pi r^2$ for both.

(Inside diameter is $3/8"$)

Area of larger circle	Area of smaller circle
$A = \pi r^2$	$A = \pi r^2$
$A = \pi(.75^2)$	$A = \pi(.1875^2)$
$A = 1.76625 \text{ in}^2$	$A = .1104466 \text{ in}^2$

Area of metal washer = larger area - smaller area

$$= 1.76625 \text{ in}^2 - .1104466 \text{ in}^2$$

$$= 1.6558034 \text{ in}^2$$

2. You have decided to fertilize your lawn. The shapes and dimensions of your lot, house, pool and garden are given in the diagram below. The shaded area will not be fertilized. If each bag of fertilizer costs $7.95 and covers 4,500 square feet, find the total number of bags needed and the total cost of the fertilizer.

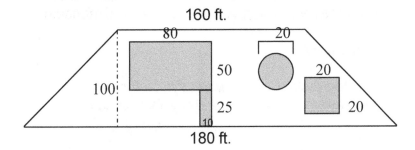

Area of Lot Area of House Area of Driveway

$A = \frac{1}{2} h(b_1 + b_2)$ $A = LW$ $A = LW$

$A = \frac{1}{2}(100)(180 + 160)$ $A = (80)(50)$ $A = (10)(25)$

$A = 17{,}000$ sq ft $A = 4{,}000$ sq ft $A = 250$ sq ft

Area of Pool Area of Garden

$A = \pi r^2$ $A = s^2$

$A = \pi(10)^2$ $A = (20)^2$

$A = 314.159$ sq. ft. $A = 400$ sq. ft.

Total area to fertilize = Lot area - (House + Driveway + Pool + Garden)
   = 17,000 - (4,000 + 250 + 314.159 + 400)
       = 12,035.841 sq ft

Number of bags needed = $\frac{\text{Total area to fertilize}}{4{,}500 \text{ sq.ft. bag}}$

$= \frac{12{,}035.841}{4{,}500}$

= 2.67 bags

Since we cannot purchase 2.67 bags we must purchase 3 full bags.

Total cost = Number of bags * $7.95
      = 3 * $7.95
      = $23.85

Examining the change in area or volume of a given figure requires first to find the existing area given the original dimensions and then finding the new area given the increased dimensions.

Sample problem:

Given the rectangle below determine the change in area if the length is increased by 5 and the width is increased by 7.

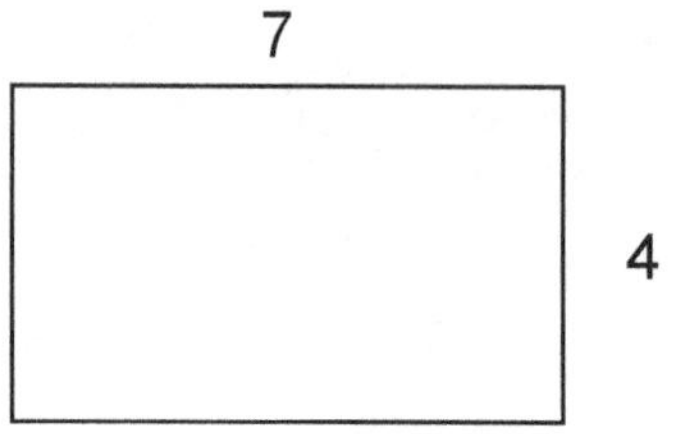

Draw and label a sketch of the new rectangle.

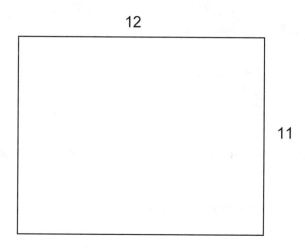

12

11

Find the areas.

Area of original = LW   Area of enlarged shape = LW
            = (7)(4)   = (12)(11)
            = 28 units^2 = 132 units^2

The change in area is 132 − 28 = 104 units^2.

Cut the **compound shape** into smaller, more familiar shapes and then compute the total area by adding the areas of the smaller parts.

Sample problem:

Find the area of the given shape.

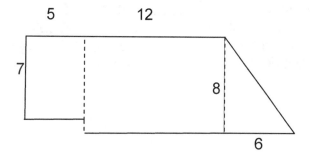

5   12

7

8

6

1. Using a dotted line we have cut the shape into smaller parts that are familiar.

2. Use the appropriate formula for each shape and find the sum of all areas.

Area 1 = LW	Area 2 = LW	Area 3 = ½bh
= (5)(7)	= (12)(8)	= ½(6)(8)
= 35 units^2	= 96 units^2	= 24 units^2

Total area = Area 1 + Area 2 + Area 3
            = 35 + 96 + 24
            = 155 units^2

It is necessary to be familiar with the metric and customary system in order to **estimate measurements**.

Some common equivalents include:

ITEM	APPROXIMATELY EQUAL TO	
	METRIC	IMPERIAL
large paper clip	1 gram	1 ounce
1 quart	1 liter	
average sized man	75 kilograms	170 pounds
1 yard	1 meter	
math textbook	1 kilogram	2 pounds
1 mile	1 kilometer	
1 foot	30 centimeters	
thickness of a dime	1 millimeter	0.1 inches

Estimate the measurement of the following items:

The length of an adult cow = _____ meters
The thickness of a compact disc = _____ millimeters
Your height = _____ meters
length of your nose = _____ centimeters
weight of your math textbook = _____ kilograms
weight of an automobile = _____ kilograms
weight of an aspirin = _____ grams

Given a set of objects and their measurements, the use of rounding procedures is helpful when attempting to round to the nearest given unit.

When rounding to a given place value, it is necessary to look at the number in the next smaller place. If this number is 5 or more, the number in the place we are rounding to is increased by one and all numbers to the right are changed to zero. If the number is less than 5, the number in the place we are rounding to stays the same and all numbers to the right are changed to zero.

One method of rounding measurements can require an additional step. First, the measurement must be converted to a decimal number. Then the rules for rounding applied.

Sample problem:

1. Round the measurements to the given units.

MEASUREMENT	ROUND TO NEAREST	ANSWER
1 foot 7 inches	foot	2 ft
5 pound 6 ounces	pound	5 pounds
5 9/16 inches	inch	6 inches

Solution:

Convert each measurement to a decimal number. Then apply the rules for rounding.

1 foot 7 inches = $1\frac{7}{12}$ ft = 1.58333 ft, round up to 2 ft

5 pounds 6 ounces = $5\frac{6}{16}$ pounds = 5.375 pound, round to 5 pounds

$5\frac{9}{16}$ inches = 5.5625 inches, round up to 6 inches

To make a **bar graph** or a **pictograph**, determine the scale to be used for the graph. Then determine the length of each bar on the graph or determine the number of pictures needed to represent each item of information. Be sure to include an explanation of the scale in the legend.

<u>Example:</u> A class had the following grades:
4 A's, 9 B's, 8 C's, 1 D, 3 F's.
Graph these on a bar graph and a pictograph.

Pictograph

Grade	Number of Students
A	☺☺☺☺
B	☺☺☺☺☺☺☺☺☺
C	☺☺☺☺☺☺☺☺
D	☺
F	☺☺☺

Bar graph

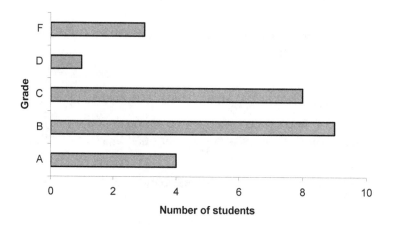

To make a **line graph**, determine appropriate scales for both the vertical and horizontal axes (based on the information to be graphed). Describe what each axis represents and mark the scale periodically on each axis. Graph the individual points of the graph and connect the points on the graph from left to right.

Example: Graph the following information using a line graph.

The number of National Merit finalists/school year

	90-'91	91-'92	92-'93	93-'94	94-'95	95-'96
Central	3	5	1	4	6	8
Wilson	4	2	3	2	3	2

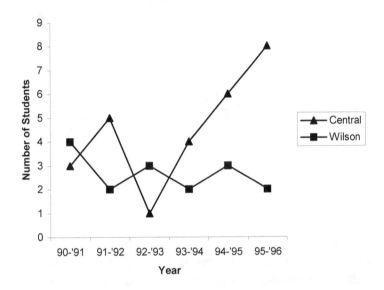

To make a **circle graph**, total all the information that is to be included on the graph. Determine the central angle to be used for each sector of the graph using the following formula:

$$\frac{\text{information}}{\text{total information}} \times 360° = \text{degrees in central} \measuredangle$$

Lay out the central angles to these sizes, label each section and include its percent.

Example: Graph this information on a circle graph:

Monthly expenses:

    Rent, $400
    Food, $150
    Utilities, $75
    Clothes, $75
    Church, $100
    Misc., $200

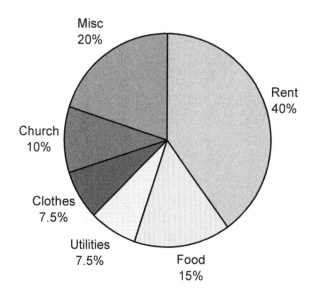

To read **a bar graph or a pictograph**, read the explanation of the scale that was used in the legend. Compare the length of each bar with the dimensions on the axes and calculate the value each bar represents. On a pictograph count the number of pictures used in the chart and calculate the value of all the pictures.

To read a circle graph, find the total of the amounts represented on the entire circle graph. To determine the actual amount that each sector of the graph represents, multiply the percent in a sector times the total amount number.

To read a chart read the row and column headings on the table. Use this information to evaluate the given information in the chart.

SUBAREA II–ALGEBRA

0004. Understand the principles and properties of the set of complex numbers and its subsets

Prime numbers are numbers that can only be factored into 1 and the number itself. When factoring into prime factors, all the factors must be numbers that cannot be factored again (without using 1). Initially numbers can be factored into any 2 factors. Check each resulting factor to see if it can be factored again. Continue factoring until all remaining factors are prime. This is the list of prime factors. Regardless of what way the original number was factored, the final list of prime factors will always be the same.

Example: Factor 30 into prime factors.

Factor 30 into any 2 factors.
$5 \cdot 6$ Now factor the 6.
$5 \cdot 2 \cdot 3$ These are all prime factors.

Factor 30 into any 2 factors.
$3 \cdot 10$ Now factor the 10.
$3 \cdot 2 \cdot 5$ These are the same prime factors even though the original factors were different.

Example: Factor 240 into prime factors.

Factor 240 into any 2 factors.
$24 \cdot 10$ Now factor both 24 and 10.
$4 \cdot 6 \cdot 2 \cdot 5$ Now factor both 4 and 6.
$2 \cdot 2 \cdot 2 \cdot 3 \cdot 2 \cdot 5$ These are prime factors.

This can also be written as $2^4 \cdot 3 \cdot 5$.

GCF is the abbreviation for the **greatest common factor**. The GCF is the largest number that is a factor of all the numbers given in a problem. The GCF can be no larger than the smallest number given in the problem. If no other number is a common factor, then the GCF will be the number 1. To find the GCF, list all possible factors of the smallest number given (include the number itself). Starting with the largest factor (which is the number itself), determine if it is also a factor of all the other given numbers. If so, that is the GCF. If that factor doesn't work, try the same method on the next smaller factor. Continue until a common factor is found. That is the GCF. Note: There can be other common factors besides the GCF.

Example: Find the GCF of 12, 20, and 36.

The smallest number in the problem is 12. The factors of 12 are 1,2,3,4,6 and 12. 12 is the largest factor, but it does not divide evenly into 20. Neither does 6, but 4 will divide into both 20 and 36 evenly. Therefore, 4 is the GCF.

Example: Find the GCF of 14 and 15.

Factors of 14 are 1,2,7 and 14. 14 is the largest factor, but it does not divide evenly into 15. Neither does 7 or 2. Therefore, the only factor common to both 14 and 15 is the number 1, the GCF.

LCM is the abbreviation for **least common multiple**. The least common multiple of a group of numbers is the smallest number that all of the given numbers will divide into. The least common multiple will always be the largest of the given numbers or a multiple of the largest number.

Example: Find the LCM of 20, 30 and 40.

The largest number given is 40, but 30 will not divide evenly into 40. The next multiple of 40 is 80 (2 x 40), but 30 will not divide evenly into 80 either. The next multiple of 40 is 120. 120 is divisible by both 20 and 30, so 120 is the LCM (least common multiple).

Example: Find the LCM of 96, 16 and 24.

The largest number is 96. 96 is divisible by both 16 and 24, so 96 is the LCM.

Divisibility Tests and Divisors

a. A number is divisible by 2 if that number is an even number (which means it ends in 0,2,4,6 or 8).

1,354 ends in 4, so it is divisible by 2. 240,685 ends in a 5, so it is not divisible by 2.

b. A number is divisible by 3 if the sum of its digits is evenly divisible by 3.

The sum of the digits of 964 is 9+6+4 = 19. Since 19 is not divisible by 3, neither is 964. The digits of 86,514 is 8+6+5+1+4 = 24. Since 24 is divisible by 3, 86,514 is also divisible by 3.

c. A number is divisible by 4 if the number in its last 2 digits is evenly divisible by 4.

The number 113,336 ends with the number 36 in the last 2 columns. Since 36 is divisible by 4, then 113,336 is also divisible by 4.

The number 135,627 ends with the number 27 in the last 2 columns. Since 27 is not evenly divisible by 4, then 135,627 is also not divisible by 4.

d. A number is divisible by 5 if the number ends in either a 5 or a 0.

225 ends with a 5 so it is divisible by 5. The number 470 is also divisible by 5 because its last digit is a 0. 2,358 is not divisible by 5 because its last digit is an 8, not a 5 or a 0.

e. A number is divisible by 6 if the number is even and the sum of its digits is evenly divisible by 3.

4,950 is an even number and its digits add to 18. (4+9+5+0 = 18) Since the number is even and the sum of its digits is 18 (which is divisible by 3), then 4950 is divisible by 6. 326 is an even number, but its digits add up to 11. Since 11 is not divisible by 3, then 326 is not divisible by 6. 698,135 is not an even number, so it cannot possibly be divided evenly by 6.

f. A number is divisible by 8 if the number in its last 3 digits is evenly divisible by 8.

The number 113,336 ends with the 3-digit number 336 in the last 3 places. Since 336 is divisible by 8, then 113,336 is also divisible by 8.

The number 465,627 ends with the number 627 in the last 3 places. Since 627 is not evenly divisible by 8, then 465,627 is also not divisible by 8.

g. A number is divisible by 9 if the sum of its digits is evenly divisible by 9.

The sum of the digits of 874 is 8+7+4 = 19. Since 19 is not divisible by 9, neither is 874. The digits of 116,514 is 1+1+6+5+1+4 = 18. Since 18 is divisible by 9, 116,514 is also divisible by 9.

h. A number is divisible by 10 if the number ends in the digit 0.

305 ends with a 5 so it is not divisible by 10. The number 2,030,270 is divisible by 10 because its last digit is a 0. 42,978 is not divisible by 10 because its last digit is an 8, not a 0.

i. Why these rules work.

All even numbers are divisible by 2 by definition. A 2-digit number (with T as the tens digit and U as the ones digit) has as its sum of he digits, T + U. Suppose this sum of T + U is divisible by 3. Then it equals 3 times some constant, K. So, T + U = 3K. Solving this for U, U = 3K - T. The original 2 digit number would be represented by 10T + U. Substituting 3K - T in place of U, this 2-digit number becomes 10T + U = 10T + (3K - T) = 9T + 3K. This 2-digit number is clearly divisible by 3, since each term is divisible by 3. Therefore, if the sum of the digits of a number is divisible by 3, then the number itself is also divisible by 3. Since 4 divides evenly into 100, 200, or 300, 4 will divide evenly into any amount of hundreds. The only part of a number that determines if 4 will divide into it evenly is the number in the last 2 places. Numbers divisible by 5 end in 5 or 0. This is clear if you look at the answers to the multiplication table for 5. Answers to the multiplication table for 6 are all even numbers. Since 6 factors into 2 times 3, the divisibility rules for 2 and 3 must both work. Any number of thousands is divisible by 8. Only the last 3 places of the number determine whether or not it is divisible by 8. A 2 digit number (with T as the tens digit and U as the ones digit) has as its sum of the digits, T + U. Suppose this sum of T + U is divisible by 9. Then it equals 9 times some constant, K. So, T + U = 9K. Solving this for U, U = 9K - T. The original 2-digit number would be represented by 10T + U. Substituting 9K - T in place of U, this 2-digit number becomes 10T + U = 10T + (9K - T) = 9T + 9K. This 2-digit number is clearly divisible by 9, since each term is divisible by 9. Therefore, if the sum of the digits of a number is divisible by 9, then the number itself is also divisible by 9. Numbers divisible by 10 must be multiples of 10 which all end in a zero.

Prime numbers are whole numbers greater than 1 that have only 2 factors, 1 and the number itself. Examples of prime numbers are 2,3,5,7,11,13,17, or 19. Note that 2 is the only even prime number.

Composite numbers are whole numbers that have more than 2 different factors. For example 9 is composite because besides factors of 1 and 9, 3 is also a factor. 70 is also composite because besides the factors of 1 and 70, the numbers 2,5,7,10,14, and 35 are also all factors.

Remember that the number 1 is neither prime nor composite.

The **exponent form** is a shortcut method to write repeated multiplication. The **base** is the factor. The **exponent** tells how many times that number is multiplied by itself.

The following are basic rules for exponents:

$a^1 = a$ for all values of a; thus $17^1 = 17$

$b^0 = 1$ for all values of b; thus $24^0 = 1$

$10^n = 1$ with n zeros; thus $10^6 = 1,000,000$

Fractions, decimals, and percents can be used interchangeably within problems.

→To change a percent into a decimal, move the decimal point two places to the left and drop off the percent sign.

→To change a decimal into a percent, move the decimal two places to the right and add on a percent sign.

→To change a fraction into a decimal, divide the numerator by the denominator.

→To change a decimal number into an equivalent fraction, write the decimal part of the number as the fraction's numerator. As the fraction's denominator use the place value of the last column of the decimal. Reduce the resulting fraction as far as possible.

<u>Example:</u> J.C. Nickels has Hunch jeans 1/4 off the usual price of $36.00. Shears and Roadkill have the same jeans 30% off their regular price of $40. Find the cheaper price.

1/4 = .25 so .25(36) = $9.00 off $36 - 9 = $27 sale price

30% = .30 so .30(40) = $12 off $40 - 12 = $28 sale price

The price at J.C Nickels is actually lower.

Define the real number system and its subsets.

a. **Natural numbers**--the counting numbers, 1,2,3,...

b. **Whole numbers**--the counting numbers along with zero, 0,1,2...

c. **Integers**--the counting numbers, their opposites, and zero, ..., ⁻1,0,1,...

d. **Rationals**--all of the fractions that can be formed from the whole numbers. Zero cannot be the denominator. In decimal form, these numbers will either be terminating or repeating decimals. Simplify square roots to determine if the number can be written as a fraction.

e. **Irrationals**--real numbers that cannot be written as a fraction. The decimal forms of these numbers are neither terminating nor repeating. <u>Examples:</u> $\pi, e, \sqrt{2}$, etc.

f. **Real numbers**--the set of numbers obtained by combining the rationals and irrationals. Complex numbers, i.e. numbers that involve i or $\sqrt{-1}$, are not real numbers.

The **Denseness Property** of real numbers states that, if all real numbers are ordered from least to greatest on a number line, there is an infinite set of real numbers between any two given numbers on the line.

Example:

Between 7.6 and 7.7, there is the rational number 7.65 in the set of real numbers.

Between 3 and 4 there exists no other natural number.

Complex numbers are of the form $a + b$ **i**, where a and b are real numbers and $i = \sqrt{-1}$. When **i** appears in an answer, it is acceptable unless it is in a denominator. When i^2 appears in a problem, it is always replaced by a $^{-}1$. Remember, $i^2 = {}^{-}1$.

To add or subtract complex numbers, add or subtract the real parts. Then add or subtract the imaginary parts and keep the **i** (just like combining like terms).

Examples: Add $(2 + 3i) + ({}^{-}7 - 4i)$.

$2 + {}^{-}7 = {}^{-}5 \qquad 3i + {}^{-}4i = {}^{-}i$ so,

$(2 + 3i) + ({}^{-}7 - 4i) = {}^{-}5 - i$

Subtract $(8 - 5i) - ({}^{-}3 + 7i)$

$8 - 5i + 3 - 7i = 11 - 12i$

To multiply 2 complex numbers, F.O.I.L. the 2 numbers together. Replace i^2 with a $^{-}1$ and finish combining like terms. Answers should have the form $a + b$ **i**.

Example: Multiply $(8 + 3i)(6 - 2i)$ F.O.I.L. this.

$48 - 16i + 18i - 6i^2 \qquad$ Let $i^2 = {}^{-}1$.

$48 - 16i + 18i - 6({}^{-}1)$

$48 - 16i + 18i + 6$

$54 + 2i \qquad\qquad$ This is the answer.

Example: Multiply $(5+8i)^2$ ← Write this out twice.

$(5+8i)(5+8i)$ F.O.I.L. this

$25+40i+40i+64i^2$ Let $i^2 = {}^-1$.

$25+40i+40i+64({}^-1)$

$25+40i+40i-64$

${}^-39+80i$ This is the answer.

When dividing 2 complex numbers, you must eliminate the complex number in the denominator. If the complex number in the denominator is of the form b **i**, multiply both the numerator and denominator by **i**. Remember to replace i^2 with **–1**, and then continue simplifying the fraction.

Example:

$$\frac{2+3i}{5i} \qquad \text{Multiply this by } \frac{i}{i}$$

$$\frac{2+3i}{5i} \times \frac{i}{i} = \frac{(2+3i)\,i}{5i\cdot i} = \frac{2i+3i^2}{5i^2} = \frac{2i+3({}^-1)}{{}^-5} = \frac{{}^-3+2i}{{}^-5} = \frac{3-2i}{5}$$

If the complex number in the denominator is of the form $a+b$ **i**, multiply both the numerator and denominator by **the conjugate of the denominator**. **The conjugate of the denominator** is the same 2 terms with the opposite sign between the 2 terms (the real term does not change signs). The conjugate of $2-3i$ is $2+3i$. The conjugate of –6 + 11i is –6 – 11i. Multiply together the factors on the top and bottom of the fraction. Remember to replace i^2 with **–1**, combine like terms, and then continue simplifying the fraction.

Example:

$$\frac{4+7i}{6-5i} \qquad \text{Multiply by } \frac{6+5i}{6+5i} \text{, the conjugate.}$$

$$\frac{(4+7i)}{(6-5i)} \times \frac{(6+5i)}{(6+5i)} = \frac{24+20i+42i+35i^2}{36+30i-30i-25i^2} = \frac{24+62i+35({}^-1)}{36-25({}^-1)} = \frac{{}^-11+62i}{61}$$

<u>Example</u>:

$$\frac{24}{^-3-5i} \qquad \text{Multiply by } \frac{^-3+5i}{^-3+5i} \text{ , the conjugate.}$$

$$\frac{24}{^-3-5i} \times \frac{^-3+5i}{^-3+5i} = \frac{^-72+120i}{9-25i^2} = \frac{^-72+120i}{9+25} = \frac{^-72+120i}{34} = \frac{^-36+60i}{17}$$

Divided everything by 2.

Vectors are used to measure displacement of an object or force.

Addition of vectors:

$$(a,b)+(c,d)=(a+c,b+d)$$

Addition Properties of vectors:

$$a+b=b+a$$
$$a+(b+c)=(a+b)+c$$
$$a+0=a$$
$$a+(^-a)=0$$

Subtraction of vectors:

$$a-b=a+(^-b) \text{ therefore,}$$
$$a-b=(a_1,a_2)+\left(^-b_1,^-b_2\right) \text{ or}$$
$$a-b=(a_1-b_1,a_2-b_2)$$

Sample problem:

If $a = \left(4, {}^-1\right)$ and $b = \left({}^-3,6\right)$, find $a+b$ and $a-b$.

Using the rule for addition of vectors:

$$\left(4,\,{}^-1\right)+\left({}^-3,6\right)=\left(4+({}^-3),\,{}^-1+6\right)$$
$$=\left(1,5\right)$$

Using the rule for subtraction of vectors:

$$\left(4,\,{}^-1\right)-\left({}^-3,6\right)=\left(4-({}^-3),\,{}^-1-6\right)$$
$$=\left(7,\,{}^-7\right)$$

0005. Understand the principles and properties of patterns and algebraic relations.

- To factor the **sum or the difference of perfect cubes**, follow this procedure:

a. Factor out any greatest common factor (GCF).

b. Make a parentheses for a binomial (2 terms) followed by a trinomial (3 terms).

c. The sign in the first parentheses is the same as the sign in the problem. The difference of cubes will have a "-" sign in the first parentheses. The sum of cubes will use a "+".

d. The first sign in the second parentheses is the opposite of the sign in the first parentheses. The second sign in the other parentheses is always a "+".

e. Determine what would be cubed to equal each term of the problem. Put those expressions in the first parentheses.

f. To make the 3 terms of the trinomial, think square - product - square. Looking at the binomial, square the first term. This is the trinomial's first term. Looking at the binomial, find the product of the two terms, ignoring the signs. This is the trinomial's second term. Looking at the binomial, square the third term. This is the trinomial's third term. Except in rare instances, the trinomial does not factor again.

Factor completely:

1.

$16x^3 + 54y^3$

$2\left(8x^3 + 27y^3\right)$ ← GCF

$2(\quad + \quad)(\quad - \quad + \quad)$ ← signs

$2(2x + 3y)(\quad - \quad + \quad)$ ← what is cubed to equal $8x^3$ or $27y^3$

$2(2x + 3y)\left(4x^2 - 6xy + 9y^2\right)$ ← square-product-square

2.

$64a^3 - 125b^3$

$(\quad - \quad)(\quad + \quad + \quad)$ ← signs

$(4a - 5b)(\quad + \quad + \quad)$ ← what is cubed to equal $64a^3$ or $125b^3$

$(4a - 5b)(16a^2 + 20ab + 25b^2)$ ← square-product-square

3.

$$27x^{27} + 343y^{12} = (3x^9 + 7y^4)(9x^{18} - 21x^9 y4 + 49y^8)$$

Note: The coefficient 27 is different from the exponent 27.

Try These:

1. $216x^3 - 125y^3$

2. $4a^3 - 32b^3$

3. $40x^{29} + 135x^2 y^3$

- To factor **a polynomial**, follow these steps:

a. **Factor out any GCF** (greatest common factor)

b. For a binomial (2 terms), check to see if the problem is the **difference of perfect squares**. If both factors are perfect squares, then it factors this way:

$$a^2 - b^2 = (a - b)(a + b)$$

If the problem is not the difference of perfect squares, then check to see if the problem is either the sum or difference of perfect cubes.

$$x^3 - 8y^3 = (x - 2y)(x^2 + 2xy + 4y^2) \qquad \leftarrow \text{difference}$$

$$64a^3 + 27b^3 = (4a + 3b)(16a^3 - 12ab + 9b^2) \qquad \leftarrow \text{sum}$$

** The sum of perfect squares does NOT factor.

c. Trinomials could be perfect squares. Trinomials can be factored into 2 binomials (un-FOILing). Be sure the terms of the trinomial are in descending order. If last sign of the trinomial is a "+", then the signs in the parentheses will be the same as the sign in front of the second term of the trinomial. If the last sign of the trinomial is a "-", then there will be one "+" and one "-" in the two parentheses. The first term of the trinomial can be factored to equal the first terms of the two factors. The last term of the trinomial can be factored to equal the last terms of the two factors. Work backwards to determine the correct factors to multiply together to get the correct center term.

Factor completely:

1. $4x^2 - 25y^2$

2. $6b^2 - 2b - 8$

3. Find a factor of $6x^2 - 5x - 4$

     a. $(3x + 2)$ b. $(3x - 2)$ c. $(6x - 1)$ d. $(2x + 1)$

The Order of Operations are to be followed when evaluating algebraic expressions. Follow these steps in order:

1. Simplify inside grouping characters such as parentheses, brackets, square root, fraction bar, etc.

2. Multiply out expressions with exponents.

3. Do multiplication or division, from left to right.

4. Do addition or subtraction, from left to right.

Samples of simplifying expressions with exponents:

$$(^-2)^3 = -8 \qquad ^-2^3 = {}^-8$$
$$(^-2)^4 = 16 \qquad ^-2^4 = 16 \qquad \text{Note change of sign.}$$
$$(\tfrac{2}{3})^3 = \tfrac{8}{27}$$
$$5^0 = 1$$
$$4^{-1} = \tfrac{1}{4}$$

- A **relation** is any set of ordered pairs.

- The **domain** of a relation is the set made of all the first coordinates of the ordered pairs.

- The **range** of a relation is the set made of all the second coordinates of the ordered pairs.

- A **function** is a relation in which different ordered pairs have different first coordinates. (No x values are repeated.)

- A **mapping** is a diagram with arrows drawn from each element of the domain to the corresponding elements of the range. If 2 arrows are drawn from the same element of the domain, then it is not a function.

- On a graph, use the **vertical line test** to look for a function. If any vertical line intersects the graph of a relation in more than one point, then the relation is not a function.

1. Determine the domain and range of this mapping.

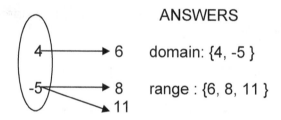

ANSWERS

domain: {4, -5 }

range : {6, 8, 11 }

2. Determine which of these are functions:

    a. $\{(1,{}^-4),(27,1)(94,5)(2,{}^-4)\}$

    b. $f(x) = 2x - 3$

    c. $A = \{(x,y) \mid xy = 24\}$

    d. $y = 3$

    e. $x = {}^-9$

    f. $\{(3,2),(7,7),(0,5),(2,{}^-4),(8,{}^-6),(1,0),(5,9),(6,{}^-4)\}$

3. Determine the domain and range of this graph.

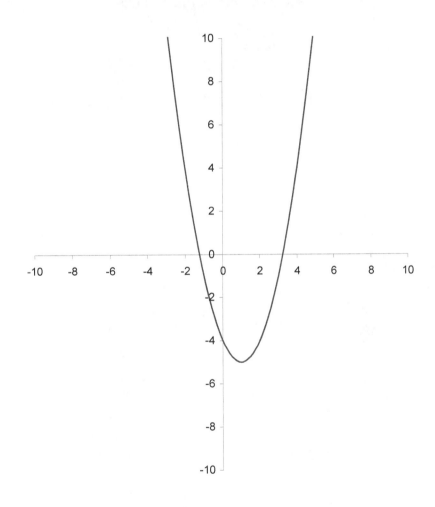

- A **relation** is any set of ordered pairs.
- The **domain** of the relation is the set of all first co-ordinates of the ordered pairs. (These are the x coordinates.)
- The **range** of the relation is the set of all second co-ordinates of the ordered pairs. (These are the y coordinates.)

1. If $A = \{(x,y) \mid y = x^2 - 6\}$, find the domain and range.

2. Give the domain and range of set B if:

$$B = \{(1, ^-2),(4, ^-2),(7, ^-2),(6, ^-2)\}$$

3. Determine the domain of this function:

$$f(x) = \frac{5x + 7}{x^2 - 4}$$

4. Determine the domain and range of these graphs.

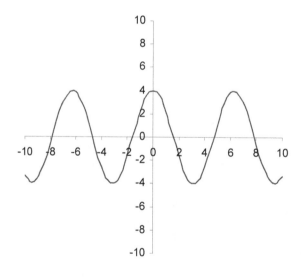

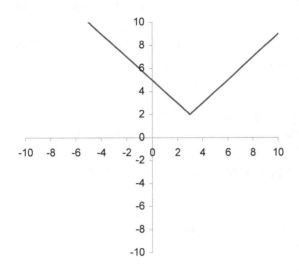

5. If $E = \{(x,y) \mid y = 5\}$, find the domain and range.

6. Determine the ordered pairs in the relation shown in this mapping.

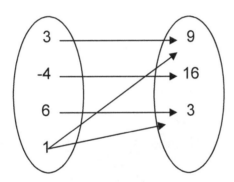

The **iterative process** involves repeated use of the same steps. A **recursive function** is an example of the iterative process. A recursive function is a function that requires the computation of all previous terms in order to find a subsequent term. Perhaps the most famous recursive function is the **Fibonacci sequence**. This is the sequence of numbers 1,1,2,3,5,8,13,21,34 … for which the next term is found by adding the previous two terms.

0006. Understand the properties of linear functions and relations

- When graphing a first-degree equation, solve for the variable. The graph of this solution will be a single point on the number line. There will be no arrows.

- When graphing a linear inequality, the dot will be hollow if the inequality sign is $<$ or $>$. If the inequality signs is either $\geq$ or $\leq$, the dot on the graph will be solid. The arrow goes to the right for $\geq$ or $>$. The arrow goes to the left for $\leq$ or $<$.

Solve:

$$5(x+2)+2x = 3(x-2)$$
$$5x+10+2x = 3x-6$$
$$7x+10 = 3x-6$$
$$4x = {}^-16$$
$$x = {}^-4$$

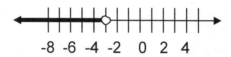

-8 -6 -4 -2   0   2   4

Solve:

$$2(3x-7) > 10x-2$$
$$6x-14 > 10x-2$$
$${}^-4x > 12$$

$x < {}^-3$ Note the change in inequality when dividint by negative numbers.

-8 -6 -4 -2   0   2   4

Solve the following equations and inequalities. Graph the solution set.

1. $5x-1 > 14$
2. $7(2x-3)+5x = 19-x$
3. $3x+42 \geq 12x-12$
4. $5-4(x+3) = 9$

- A first degree equation has an equation of the form $ax + by = c$. To graph this equation, find either one point and the slope of the line or find two points. To find a point and slope, solve the equation for y. This gets the equation in **slope intercept form**, $y = mx + b$. The point (0,b) is the y-intercept and m is the line's slope. To find any 2 points, substitute any 2 numbers for x and solve for y. To find the intercepts, substitute 0 for x and then 0 for y.

- Remember that graphs will go up as they go to the right when the slope is positive. Negative slopes make the lines go down as they go to the right.

- If the equation solves to **x = any number**, then the graph is a **vertical line**.

- If the equation solves to **y = any number**, then the graph is a **horizontal line**.

- When graphing a linear inequality, the line will be dotted if the inequality sign is < or >. If the inequality signs are either ≥ or ≤, the line on the graph will be a solid line. Shade above the line when the inequality sign is ≥ or >. Shade below the line when the inequality sign is < or ≤. Inequalities of the form $x >, x ≤, x <,$ or $x ≥$ number, draw a vertical line (solid or dotted). Shade to the right for > or ≥. Shade to the left for < or ≤. Remember: **Dividing or multiplying by a negative number will reverse the direction of the inequality sign.**

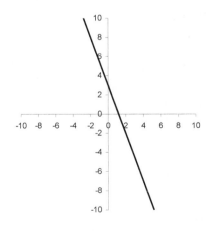

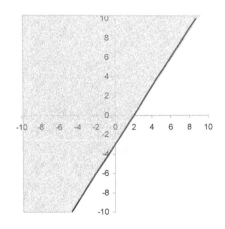

$$5x + 2y = 6$$
$$y = {}^-5/2x + 3$$

$$3x - 2y ≥ 6$$
$$y ≤ 3/2x - 3$$

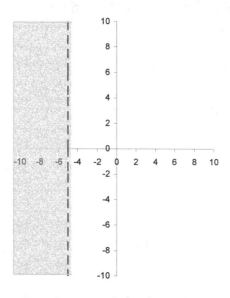

$$3x + 12 < -3$$

$$x < {}^-5$$

Graph the following:

1. $2x - y = {}^-4$
2. $x + 3y > 6$
3. $3x + 2y \leq 2y - 6$

- A first degree equation has an equation of the form $ax + by = c$. To find the slope of a line, solve the equation for y. This gets the equation into **slope intercept form**, $y = mx + b$. m is the line's slope.

- To find the y intercept, substitute 0 for x and solve for y. This is the y intercept. The y intercept is also the value of b in $y = mx + b$.

- To find the x intercept, substitute 0 for y and solve for x. This is the x intercept.

- If the equation solves to **x = any number**, then the graph is a **vertical line**. It only has an x intercept. Its slope is **undefined**.

- If the equation solves to **y = any number**, then the graph is a **horizontal line**. It only has a y intercept. Its slope is 0 (zero).

1. Find the slope and intercepts of $3x + 2y = 14$.

$$3x + 2y = 14$$
$$2y = {}^-3x + 14$$
$$y = {}^-3/2\ x + 7$$

The slope of the line is $^-3/2$, the value of m.
The y intercept of the line is 7.

The intercepts can also be found by substituting 0 in place of the other variable in the equation.

To find the y intercept:
let $x = 0$; $3(0) + 2y = 14$
$0 + 2y = 14$
$2y = 14$
$y = 7$
$(0,7)$ is the y intercept.

To find the x intercept:
let $y = 0$; $3x + 2(0) = 14$
$3x + 0 = 14$
$3x = 14$
$x = 14/3$
$(14/3, 0)$ is the x intercept.

Find the slope and the intercepts (if they exist) for these equations:

1. $5x + 7y = {}^-70$
2. $x - 2y = 14$
3. $5x + 3y = 3(5 + y)$
4. $2x + 5y = 15$

- The **equation of a graph** can be found by finding its slope and its y intercept. To find the slope, find 2 points on the graph where co-ordinates are integer values. Using points: (x_1, y_1) and (x_2, y_2).

$$slope = \frac{y_2 - y_1}{x_2 - x_1}$$

The y intercept is the y coordinate of the point where a line crosses the y axis. The equation can be written in slope-intercept form, which is $y = mx + b$, where m is the slope and b is the y intercept. To rewrite the equation into some other form, multiply each term by the common denominator of all the fractions. Then rearrange terms as necessary.

- If the graph is a **vertical line**, then the equation solves to x = the x co-ordinate of any point on the line.

- If the graph is a **horizontal line**, then the equation solves to y = the y coordinate of any point on the line.

- Given two points on a line, the first thing to do is to find the slope of the line. If 2 points on the graph are (x_1, y_1) and (x_2, y_2), then the slope is found using the formula:

$$\text{slope} = \frac{y_2 - y_1}{x_2 - x_1}$$

The slope will now be denoted by the letter **m**. To write the equation of a line, choose either point. Substitute them into the formula:

$$Y - y_a = m(X - x_a)$$

Remember (x_a, y_a) can be (x_1, y_1) or (x_2, y_2) If **m**, the value of the slope, is distributed through the parentheses, the equation can be rewritten into other forms of the equation of a line.

Find the equation of a line through $(9, ^-6)$ and $(^-1, 2)$.

$$\text{slope} = \frac{y_2 - y_1}{x_2 - x_1} = \frac{2 - ^-6}{^-1 - 9} = \frac{8}{^-10} = -\frac{4}{5}$$

$$Y - y_a = m(X - x_a) \rightarrow Y - 2 = ^-4/5(X - ^-1) \rightarrow$$
$$Y - 2 = ^-4/5(X + 1) \rightarrow Y - 2 = ^-4/5 X - 4/5 \rightarrow$$
$$Y = ^-4/5 \ X + 6/5 \quad \text{This is the slope-intercept form.}$$

Multiplying by 5 to eliminate fractions, it is:

$$5Y = ^-4X + 6 \rightarrow 4X + 5Y = 6 \quad \text{Standard form.}$$

Write the equation of a line through these two points:
1. $(5,8)$ and $(^-3,2)$
2. $(11,10)$ and $(11, ^-3)$
3. $(^-4,6)$ and $(6,12)$
4. $(7,5)$ and $(^-3,5)$

When given the following system of equations:

$$ax + by = e$$
$$cx + dy = f$$

the matrix equation is written in the form:

$$\begin{pmatrix} a & b \\ c & d \end{pmatrix}\begin{pmatrix} x \\ y \end{pmatrix} = \begin{pmatrix} e \\ f \end{pmatrix}$$

The solution is found using the inverse of the matrix of coefficients. Inverse of matrices can be written as follows:

$$A^{-1} = \frac{1}{\text{determinant of } A}\begin{pmatrix} d & {}^-b \\ {}^-c & a \end{pmatrix}$$

Sample Problem:
1. Write the matrix equation of the system.

$$3x - 4y = 2$$
$$2x + y = 5$$

$$\begin{pmatrix} 3 & {}^-4 \\ 2 & 1 \end{pmatrix}\begin{pmatrix} x \\ y \end{pmatrix} = \begin{pmatrix} 2 \\ 5 \end{pmatrix}$$ Definition of matrix equation.

$$\begin{pmatrix} x \\ y \end{pmatrix} = \frac{1}{11}\begin{pmatrix} 1 & 4 \\ {}^-2 & 3 \end{pmatrix}\begin{pmatrix} 2 \\ 5 \end{pmatrix}$$ Multiply by the inverse of the coefficient matrix.

$$\begin{pmatrix} x \\ y \end{pmatrix} = \frac{1}{11}\begin{pmatrix} 22 \\ 11 \end{pmatrix}$$ Matrix multiplication.

$$\begin{pmatrix} x \\ y \end{pmatrix} = \begin{pmatrix} 2 \\ 1 \end{pmatrix}$$ Scalar multiplication. The solution is (2,1).

Practice problems:

1.
$$x + 2y = 5$$
$$3x + 5y = 14$$

2.
$${}^-3x + 4y - z = 3$$
$$x + 2y - 3z = 9$$
$$y - 5z = {}^-1$$

Word problems can sometimes be solved by using a system of two equations in 2 unknowns. This system can then be solved using **substitution**, the **addition-subtraction method**, or **determinants**.

Example: Farmer Greenjeans bought 4 cows and 6 sheep for $1700. Mr. Ziffel bought 3 cows and 12 sheep for $2400. If all the cows were the same price and all the sheep were another price, find the price charged for a cow or for a sheep.

Let x = price of a cow
Let y = price of a sheep

Then Farmer Greenjeans' equation would be: $4x + 6y = 1700$
Mr. Ziffel's equation would be: $3x + 12y = 2400$

To solve by **addition-subtraction**:

Multiply the first equation by $^-2$: $^-2(4x + 6y = 1700)$
Keep the other equation the same : $(3x + 12y = 2400)$
By doing this, the equations can be added to each other to eliminate one variable and solve for the other variable.

$$^-8x - 12y = {}^-3400$$
$$3x + 12y = 2400 \qquad \text{Add these equations.}$$
$$^-5x \qquad = {}^-1000$$

$x = 200 \leftarrow$ the price of a cow was $200.
Solving for y , $y = 150 \leftarrow$ the price of a sheep,$150.

To solve by **substitution**:
Solve one of the equations for a variable. (Try to make an equation without fractions if possible.) Substitute this expression into the equation that you have not yet used. Solve the resulting equation for the value of the remaining variable.

$$4x + 6y = 1700$$
$$3x + 12y = 2400 \leftarrow \text{Solve this equation for } x.$$

It becomes $x = 800 - 4y$. Now substitute $800 - 4y$ in place of x in the OTHER equation. $4x + 6y = 1700$ now becomes:

$$4(800 - 4y) + 6y = 1700$$
$$3200 - 16y + 6y = 1700$$
$$3200 - 10y = 1700$$
$$^-10y = {}^-1500$$
$$y = 150, \text{ or } \$150 \text{ for a sheep.}$$

Substituting 150 back into an equation for y, find x.

$$4x + 6(150) = 1700$$
$$4x + 900 = 1700$$
$$4x = 800 \text{ so } x = 200 \text{ for a cow.}$$

To solve by **determinants**:

Let x = price of a cow
Let y = price of a sheep

Then Farmer Greenjeans' equation would be: $4x + 6y = 1700$
Mr. Ziffel's equation would be: $3x + 12y = 2400$

To solve this system using determinants, make one 2 by 2 determinant divided by another 2 by 2 determinant. The bottom determinant is filled with the x and y term coefficients. The top determinant is almost the same as this bottom determinant. The only difference is that when you are solving for x, the x coefficients are replaced with the constants. Likewise, when you are solving for y, the y coefficients are replaced with the constants. To find the value of a 2 by 2 determinant,

$\begin{pmatrix} a & b \\ c & d \end{pmatrix}$, is found by $ad - bc$.

$$x = \frac{\begin{pmatrix} 1700 & 6 \\ 2400 & 12 \end{pmatrix}}{\begin{pmatrix} 4 & 6 \\ 3 & 12 \end{pmatrix}} = \frac{1700(12) - 6(2400)}{4(12) - 6(3)} = \frac{20400 - 14400}{48 - 18} = \frac{6000}{30} = 200$$

$$y = \frac{\begin{pmatrix} 4 & 1700 \\ 3 & 2400 \end{pmatrix}}{\begin{pmatrix} 4 & 6 \\ 3 & 12 \end{pmatrix}} = \frac{2400(4) - 3(1700)}{4(12) - 6(3)} = \frac{9600 - 5100}{48 - 18} = \frac{4500}{30} = 150$$

NOTE: The bottom determinant is always the same value for each   letter.

Word problems can sometimes be solved by using a system of three equations in 3 unknowns. This system can then be solved using **substitution**, the **addition-subtraction method**, or **determinants**.

To solve by **substitution**:

Example: Mrs. Allison bought 1 pound of potato chips, a 2 pound beef roast, and 3 pounds of apples for a total of $ 8.19. Mr. Bromberg bought a 3 pound beef roast and 2 pounds of apples for $ 9.05. Kathleen Kaufman bought 2 pounds of potato chips, a 3 pound beef roast, and 5 pounds of apples for $ 13.25. Find the per pound price of each item.

Let x = price of a pound of potato chips
Let y = price of a pound of roast beef
Let z = price of a pound of apples

Mrs. Allison's equation would be: $1x + 2y + 3z = 8.19$
Mr. Bromberg's equation would be: $3y + 2z = 9.05$
K. Kaufman's equation would be: $2x + 3y + 5z = 13.25$

Take the first equation and solve it for x. (This was chosen because x is the easiest variable to get alone in this set of equations.) This equation would become:

$$x = 8.19 - 2y - 3z$$

Substitute this expression into the other equations in place of the letter x:

$$3y + 2z = 9.05 \leftarrow \text{equation 2}$$
$$2(8.19 - 2y - 3z) + 3y + 5z = 13.25 \leftarrow \text{equation 3}$$

Simplify the equation by combining like terms:

$$3y + 2z = 9.05 \leftarrow \text{equation 2}$$
$$* \ ^-1y - 1z = {}^-3.13 \leftarrow \text{equation 3}$$

Solve equation 3 for either y or z:

$$y = 3.13 - z \quad \text{Substitute this into equation 2 for } y:$$

$$3(3.13 - z) + 2z = 9.05 \leftarrow \text{equation 2}$$
$$^-1y - 1z = {}^-3.13 \leftarrow \text{equation 3}$$

Combine like terms in equation 2:

$$9.39 - 3z + 2z = 9.05$$
$$z = .34 \quad \text{per pound price of apples}$$

Substitute .34 for z in the starred equation above to solve for y:
$$y = 3.13 - z \text{ becomes } y = 3.13 - .34, \text{ so}$$
$$y = 2.79 = \text{per pound price of roast beef}$$

Substituting .34 for z and 2.79 for y in one of the original equations, solve for x:

$$1x + 2y + 3z = 8.19$$
$$1x + 2(2.79) + 3(.34) = 8.19$$
$$x + 5.58 + 1.02 = 8.19$$
$$x + 6.60 = 8.19$$
$$x = 1.59 \quad \text{per pound of potato chips}$$

$$(x, y, z) = (\ 1.59,\ 2.79,\ .34)$$

To solve by **addition-subtraction**:

Choose a letter to eliminate. Since the second equation is already missing an x, let's eliminate x from equations 1 and 3.

1) $1x + 2y + 3x = 8.19 \leftarrow$ Multiply by $^-2$ below.
2) $3y + 2z = 9.05$
3) $2x + 3y + 5z = 13.25$

$^-2(1x + 2y + 3z = 8.19)$ $=$ $^-2x - 4y - 6z = ^-16.38$
Keep equation 3 the same : $\underline{2x + 3y + 5z = 13.25}$

By doing this, the equations $^-y - z = ^-3.13 \leftarrow$ equation 4
can be added to each other to
eliminate one variable.

The equations left to solve are equations 2 and 4:

   $^-y - z = ^-3.13 \leftarrow$ equation 4
   $3y + 2z = 9.05 \leftarrow$ equation 2

Multiply equation 4 by 3: $3(^-y - z = ^-3.13)$
Keep equation 2 the same: $3y + 2z = 9.05$

         $^-3y - 3z = ^-9.39$
         $\underline{3y + 2z = 9.05}$ Add these equations.
         $^-1z = ^-.34$
             $z = .34 \leftarrow$ the per pound price of apples
solving for y, $y = 2.79 \leftarrow$ the per pound roast beef price
solving for x, $x = 1.59 \leftarrow$ potato chips, per pound price

To solve by **substitution**:

Solve one of the 3 equations for a variable. (Try to make an equation without fractions if possible.) Substitute this expression into the other 2 equations that you have not yet used.

1) $1x + 2y + 3z = 8.19$ ← Solve for x.
2) $3y + 2z = 9.05$
3) $2x + 3y + 5z = 13.25$

   Equation 1 becomes $x = 8.19 - 2y - 3z$.

Substituting this into equations 2 and 3, they become:

2) $3y + 2z = 9.05$
3) $2(8.19 - 2y - 3z) + 3y + 5z = 13.25$
   $16.38 - 4y - 6z + 3y + 5z = 13.25$

   $^-y - z = {^-}3.13$

The equations left to solve are :

   $3y + 2z = 9.05$

   $^-y - z = {^-}3.13$ ← Solve for either y or z.

It becomes $y = 3.13 - z$. Now substitute $3.13 - z$ in place of y in the OTHER equation. $3y + 2z = 9.05$ now becomes:

$$3(3.13 - z) + 2z = 9.05$$
$$9.39 - 3z + 2z = 9.05$$
$$9.39 - z = 9.05$$

$$^-z = {^-}.34$$
$$z = .34, \text{ or } \$.34/\text{lb of apples}$$

Substituting .34 back into an equation for z, find y.
   $3y + 2z = 9.05$
   $3y + 2(.34) = 9.05$
   $3y + .68 = 9.05$ so $y = 2.79/\text{lb of roast beef}$

Substituting .34 for z and 2.79 for y into one of the original equations, it becomes:
   $2x + 3y + 5z = 13.25$
   $2x + 3(2.79) + 5(.34) = 13.25$
   $2x + 8.37 + 1.70 = 13.25$
   $2x + 10.07 = 13.25$, so $x = 1.59/\text{lb of potato chips}$

To solve by **determinants**:

Let x = price of a pound of potato chips
Let y = price of a pound of roast beef
Let z = price of a pound of apples

1) $1x + 2y + 3z = 8.19$
2) $3y + 2z = 9.05$
3) $2x + 3y + 5z = 13.25$

To solve this system using determinants, make one 3 by 3 determinant divided by another 3 by 3 determinant. The bottom determinant is filled with the x, y, and z term coefficients. The top determinant is almost the same as this bottom determinant. The only difference is that when you are solving for x, the x coefficients are replaced with the constants. When you are solving for y, the y coefficients are replaced with the constants. Likewise, when you are solving for z, the z coefficients are replaced with the constants. To find the value of a 3 by 3 determinant,

$$\begin{pmatrix} a & b & c \\ d & e & f \\ g & h & i \end{pmatrix}$$ is found by the following steps:

Copy the first two columns to the right of the determinant:

$$\begin{pmatrix} a & b & c \\ d & e & f \\ g & h & i \end{pmatrix} \begin{matrix} a & b \\ d & e \\ g & h \end{matrix}$$

Multiply the diagonals from top left to bottom right, and add these diagonals together.

$$\begin{pmatrix} a^* & b^\circ & c^\bullet \\ d & e^* & f^\circ \\ g & h & i^* \end{pmatrix} \begin{matrix} a & b \\ d^\bullet & e \\ g^\circ & h^\bullet \end{matrix} = a^*e^*i^* + b^\circ f^\circ g^\circ + c^\bullet d^\bullet h^\bullet$$

Then multiply the diagonals from bottom left to top right, and add these diagonals together.

$$\begin{pmatrix} a & b & c^* \\ d & e^* & f^\circ \\ g^* & h^\circ & i^\bullet \end{pmatrix} \begin{matrix} a^\circ & b^\bullet \\ d^\bullet & e \\ g & h \end{matrix} = g^* e^* c^* + h^\circ f^\circ a^\circ + i^\bullet d^\bullet b^\bullet$$

Subtract the first diagonal total minus the second diagonal total:

$$(= a^* e^* i^* + b^\circ f^\circ g^\circ + c^\bullet d^\bullet h^\bullet) - (= g^* e^* c^* + h^\circ f^\circ a^\circ + i^\bullet d^\bullet b^\bullet)$$

This gives the value of the determinant. To find the value of a variable, divide the value of the top determinant by the value of the bottom determinant.

1) $1x + 2y + 3z = 8.19$
2) $3y + 2z = 9.05$
3) $2x + 3y + 5z = 13.25$

$$x = \frac{\begin{pmatrix} 8.19 & 2 & 3 \\ 9.05 & 3 & 2 \\ 13.25 & 3 & 5 \end{pmatrix}}{\begin{pmatrix} 1 & 2 & 3 \\ 0 & 3 & 2 \\ 2 & 3 & 5 \end{pmatrix}}$$ solve each determinant using the method shown below

Multiply the diagonals from top left to bottom right, and add these diagonals together.

$$\begin{pmatrix} 8.19^* & 2^\circ & 3^\bullet \\ 9.05 & 3^* & 2^\circ \\ 13.25 & 3 & 5^* \end{pmatrix} \begin{matrix} 8.19 & 2 \\ 9.05^\bullet & 3 \\ 13.25^\circ & 3^\bullet \end{matrix}$$

$$= (8.19^*)(3^*)(5^*) + (2^\circ)(2^\circ)(13.25^\circ) + (3^\bullet)(9.05^\bullet)(3^\bullet)$$

Then multiply the diagonals from bottom left to top right, and add these diagonals together.

$$\begin{pmatrix} 8.19 & 2 & 3^* \\ 9.05 & 3^* & 2^\circ \\ 13.25^* & 3^\circ & 5^\bullet \end{pmatrix} \begin{array}{cc} 8.19^\circ & 2^\bullet \\ 9.05^\bullet & 3 \\ 13.25 & 3 \end{array}$$

$$= (13.25^*)(3^*)(3^*) + (3^\circ)(2^\circ)(8.19^\circ) + (5^\bullet)(9.05^\bullet)(2^\bullet)$$

Subtract the first diagonal total minus the second diagonal total:

$$(8.19^*)(3^*)(5^*) + (2^\circ)(2^\circ)(13.25^\circ) + (3^\bullet)(9.05^\bullet)(3^\bullet) = \ 257.30$$
$$- \ (13.25^*)(3^*)(3^*) + (3^\circ)(2^\circ)(8.19^\circ) + (5^\bullet)(9.05^\bullet)(2^\bullet) = ^- 258.89$$
$$^-1.59$$

Do same multiplying and subtraction procedure for the bottom determinant to get $^-1$ as an answer. Now divide:

$$\frac{^-1.59}{^-1} = \$1.59/\text{lb of potato chips}$$

$$y = \frac{\begin{pmatrix} 1 & 8.19 & 3 \\ 0 & 9.05 & 2 \\ 2 & 13.25 & 5 \end{pmatrix}}{\begin{pmatrix} 1 & 2 & 3 \\ 0 & 3 & 2 \\ 2 & 3 & 5 \end{pmatrix}} = \frac{^-2.79}{^-1} = \$2.79/\text{lb of roast beef}$$

NOTE: The bottom determinant is always the same value for each letter.

$$z = \frac{\begin{pmatrix} 1 & 2 & 8.19 \\ 0 & 3 & 9.05 \\ 2 & 3 & 13.25 \end{pmatrix}}{\begin{pmatrix} 1 & 2 & 3 \\ 0 & 3 & 2 \\ 2 & 3 & 5 \end{pmatrix}} = \frac{^-.34}{^-1} = \$.34/\text{lb of apples}$$

59

To graph an inequality, solve the inequality for y. This gets the inequality in **slope intercept form**, (for example : $y < mx + b$). The point (0,b) is the y-intercept and m is the line's slope.

- If the inequality solves to $x \geq$ **any number**, then the graph includes a **vertical line**.

- If the inequality solves to $y \leq$ **any number**, then the graph includes a **horizontal line**.

- When graphing a linear inequality, the line will be dotted if the inequality sign is < or >. If the inequality signs are either $\geq$ or $\leq$, the line on the graph will be a solid line. Shade above the line when the inequality sign is $\geq$ or >. Shade below the line when the inequality sign is < or $\leq$. For inequalities of the forms $x >$ number, $x \leq$ number , $x <$ number ,or $x \geq$ number, draw a vertical line (solid or dotted). Shade to the right for > or $\geq$. Shade to the left for < or $\leq$.

Remember: **Dividing or multiplying by a negative number will reverse the direction of the inequality sign.**

Use these rules to graph and shade each inequality. The solution to a system of linear inequalities consists of the part of the graph that is shaded for each inequality. For instance, if the graph of one inequality was shaded with red, and the graph of another inequality was shaded with blue, then the overlapping area would be shaded purple. The purple area would be the points in the solution set of this system.

Example: Solve by graphing:

$$x + y \leq 6$$
$$x - 2y \leq 6$$

Solving the inequalities for y, they become:

$y \leq {}^-x + 6$ (y intercept of 6 and slope = $^-1$)
$y \geq 1/2\,x - 3$ (y intercept of $^-3$ and slope = $1/2$)

A graph with shading is shown below:

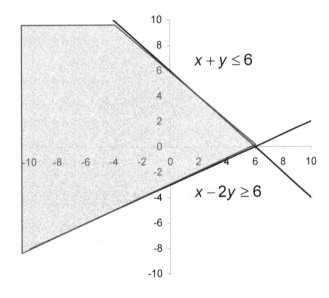

0007. Understand the properties of quadratic and higher-order polynomial and relations.

A **quadratic equation** is written in the form $ax^2 + bx + c = 0$. To solve a quadratic equation by factoring, at least one of the factors must equal zero.

Example:
Solve the equation.

$x^2 + 10x - 24 = 0$

$(x + 12)(x - 2) = 0$ Factor.

$x + 12 = 0$ or $x - 2 = 0$ Set each factor equal to 0.

$x = {}^-12$ $x = 2$ Solve.

Check:

$x^2 + 10x - 24 = 0$

$({}^-12)^2 + 10({}^-12) - 24 = 0$ $(2)^2 + 10(2) - 24 = 0$

$144 - 120 - 24 = 0$ $4 + 20 - 24 = 0$

$0 = 0$ $0 = 0$

A quadratic equation that cannot be solved by factoring can be solved by **completing the square**.

Example:

Solve the equation.

$x^2 - 6x + 8 = 0$

$x^2 - 6x = {}^-8$ Move the constant to the right side.

$x^2 - 6x + 9 = {}^-8 + 9$ Add the square of half the coefficient of x to both sides.

$(x - 3)^2 = 1$ Write the left side as a perfect square.

$x - 3 = \pm\sqrt{1}$ Take the square root of both sides.

$x - 3 = 1$ $x - 3 = {}^-1$ Solve.

$x = 4$ $x = 2$

Check:

$x^2 - 6x + 8 = 0$

$4^2 - 6(4) + 8 = 0$ $2^2 - 6(2) + 8 = 0$

$16 - 24 + 8 = 0$ $4 - 12 + 8 = 0$

$0 = 0$ $0 = 0$

The general technique for **graphing quadratics** is the same as for graphing linear equations. Graphing quadratic equations, however, results in a parabola instead of a straight line.

<u>Example</u>:

Graph $y = 3x^2 + x - 2$.

x	$y = 3x^2 + x - 2$
$^-2$	8
$^-1$	0
0	$^-2$
1	2
2	12

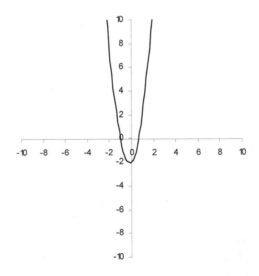

To solve **a quadratic equation using the quadratic formula**, be sure that your equation is in the form $ax^2 + bx + c = 0$. Substitute these values into the formula:

$$x = \frac{-b \pm \sqrt{b^2 - 4ac}}{2a}$$

Example:

Solve the equation.

$$3x^2 = 7 + 2x \rightarrow 3x^2 - 2x - 7 = 0$$

$$a = 3 \quad b = {}^-2 \quad c = {}^-7$$

$$x = \frac{-({}^-2) \pm \sqrt{({}^-2)^2 - 4(3)({}^-7)}}{2(3)}$$

$$x = \frac{2 \pm \sqrt{4 + 84}}{6}$$

$$x = \frac{2 \pm \sqrt{88}}{6}$$

$$x = \frac{2 \pm 2\sqrt{22}}{6}$$

$$x = \frac{1 \pm \sqrt{22}}{3}$$

Some **word problems** will give a quadratic equation to be solved. When the quadratic equation is found, set it equal to zero and solve the equation by factoring or the quadratic formula. Examples of this type of problem follow.

Example:
Alberta (A) is a certain distance north of Boston (B). The distance from Boston east to Carlisle (C) is 5 miles more than the distance from Alberta to Boston. The distance from Alberta to Carlisle is 10 miles more than the distance from Alberta to Boston. How far is Alberta from Carlisle?

Solution:
Since north and east form a right angle, these distances are the lengths of the legs of a right triangle. If the distance from Alberta to Boston is x, then from Boston to Carlisle is $x + 5$, and the distance from Alberta to Carlisle is $x + 10$.

The equation is: $AB^2 + BC^2 = AC^2$

$$x^2 + (x+5)^2 = (x+10)^2$$

$$x^2 + x^2 + 10x + 25 = x^2 + 20x + 100$$

$$2x^2 + 10x + 25 = x^2 + 20x + 100$$

$$x^2 - 10x - 75 = 0$$

$(x-15)(x+5) = 0$ Distance cannot be negative.

$x = 15$ Distance from Alberta to Boston.

$x + 5 = 20$ Distance from Boston to Carlisle.

$x + 10 = 25$ Distance from Alberta to Carlisle.

<u>Example</u>:
The square of a number is equal to 6 more than the original number. Find the original number.

Solution: If x = original number, then the equation is:

$x^2 = 6 + x$ Set this equal to zero.

$x^2 - x - 6 = 0$ Now factor.

$(x-3)(x+2) = 0$

$x = 3$ or $x = {}^- 2$ There are 2 solutions, 3 or $^-2$.

Try these:

1. One side of a right triangle is 1 less than twice the shortest side, while the third side of the triangle is 1 more than twice the shortest side. Find all 3 sides.

2. Twice the square of a number equals 2 less than 5 times the number. Find the number(s).

The **discriminant of a quadratic equation** is the part of the quadratic formula that is usually inside the radical sign, $b^2 - 4ac$.

$$x = \frac{-b \pm \sqrt{b^2 - 4ac}}{2a}$$

The radical sign is NOT part of the discriminant!! Determine the value of the discriminant by substituting the values of a, b, and c from $ax^2 + bx + c = 0$.

-If the value of the discriminant is **any negative number**, then there are **two complex roots** including "i".
-If the value of the discriminant is **zero**, then there is only **1 real rational root**. This would be a double root.
-If the value of the discriminant is **any positive number that is also a perfect square**, then there are **two real rational roots.** (There are no longer any radical signs.)
-If the value of the discriminant is **any positive number that is NOT a perfect square**, then there are **two real irrational roots.** (There are still unsimplified radical signs.)

Example:

Find the value of the discriminant for the following equations. Then determine the number and nature of the solutions of that quadratic equation.

$2x^2 - 5x + 6 = 0$
$a = 2$, $b = {}^-5$, $c = 6$ so $b^2 - 4ac = ({}^-5)^2 - 4(2)(6) = 25 - 48 = {}^-23$.

Since ${}^-23$ is a negative number, there are **two complex roots** including "i".

$3x^2 - 12x + 12 = 0$
$a = 3$, $b = {}^-12$, $c = 12$ so $b^2 - 4ac = ({}^-12)^2 - 4(3)(12) = 144 - 144 = 0$
.

Since 0 is the value of the discriminant, there is only **1 real rational root**.

$6x^2 - x - 2 = 0$
$a = 6$, $b = {}^-1$, $c = {}^-2$ so $b^2 - 4ac = ({}^-1)^2 - 4(6)({}^-2) = 1 + 48 = 49$.

Since 49 is positive and is also a perfect square $(\sqrt{49}) = 7$, then there are **two real rational roots.**

Try these:

1. $6x^2 - 7x - 8 = 0$

2. $10x^2 - x - 2 = 0$

3. $25x^2 - 80x + 64 = 0$

Follow these steps to write a quadratic equation from its roots:

1. Add the roots together. The answer is their **sum**. Multiply the roots together. The answer is their **product**.
2. A quadratic equation can be written using the sum and product like this:

$$x^2 + (\text{opposite of the sum})x + \text{product} = 0$$

3. If there are any fractions in the equation, multiply every term by the common denominator to eliminate the fractions. This is the quadratic equation.
4. If a quadratic equation has only 1 root, use it twice and follow the first 3 steps above.

Example:
Find a quadratic equation with roots of 4 and ⁻9.

Solutions:
The sum of 4 and ⁻9 is ⁻5. The product of 4 and ⁻9 is ⁻36. The equation would be:

$$x^2 + (\text{opposite of the sum})x + \text{product} = 0$$
$$x^2 + 5x - 36 = 0$$

Find a quadratic equation with roots of $5 + 2i$ and $5 - 2i$.

Solutions:
The sum of $5 + 2i$ and $5 - 2i$ is 10. The product of $5 + 2i$ and $5 - 2i$ is $25 - 4i^2 = 25 + 4 = 29$.

The equation would be:

$$x^2 + (\text{opposite of the sum})x + \text{product} = 0$$
$$x^2 - 10x + 29 = 0$$

Find a quadratic equation with roots of $2/3$ and $^-3/4$.

Solutions:

The sum of $2/3$ and $^-3/4$ is $^-1/12$. The product of $2/3$ and $^-3/4$ is $^-1/2$.

The equation would be :

$$x^2 + \text{(opposite of the sum)}x + \text{product} = 0$$
$$x^2 + 1/12\,x - 1/2 = 0$$

Common denominator = 12, so multiply by 12.

$$12(x^2 + 1/12\,x - 1/2 = 0$$
$$12x^2 + 1x - 6 = 0$$
$$12x^2 + x - 6 = 0$$

Try these:

1. Find a quadratic equation with a root of 5.
2. Find a quadratic equation with roots of $8/5$ and $^-6/5$.
3. Find a quadratic equation with roots of 12 and $^-3$.

Some word problems can be solved by setting up a quadratic equation or inequality. Examples of this type could be problems that deal with finding a maximum area. Examples follow:

Example 1:
A family wants to enclose 3 sides of a rectangular garden with 200 feet of fence. In order to have a garden with an area of **at least** 4800 square feet, find the dimensions of the garden. Assume that the fourth side of the garden is already bordered by a wall or a fence.

Existing Wall

Solution:
Let $x =$ distance
from the wall

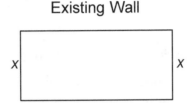

Then 2x feet of fence is used for these 2 sides. The remaining side of the garden would use the rest of the 200 feet of fence, that is, $200 - 2x$ feet of fence. Therefore the width of the garden is x feet and the length is $200 - 2x$ ft.

The area, $200x - 2x^2$, needs to be greater than or equal to 4800 sq. ft. So, this problem uses the inequality $4800 \leq 200x - 2x^2$. This becomes $2x^2 - 200x + 4800 \leq 0$. Solving this, we get:

$$200x - 2x^2 \geq 4800$$
$$-2x^2 + 200x - 4800 \geq 0$$
$$2\left(-x^2 + 100x - 2400\right) \geq 0$$
$$-x^2 + 100x - 2400 \geq 0$$
$$(-x + 60)(x - 40) \geq 0$$
$$-x + 60 \geq 0$$
$$-x \geq -60$$
$$x \leq 60$$
$$x - 40 \geq 0$$
$$x \geq 40$$

So the area will be at least 4800 square feet if the width of the garden is from 40 up to 60 feet.

Quadratic equations can be used to model different real life situations. The graphs of these quadratics can be used to determine information about this real life situation.

Example:
The height of a projectile fired upward at a velocity of v meters per second from an original height of h meters is $y = h + vx - 4.9x^2$. If a rocket is fired from an original height of 250 meters with an original velocity of 4800 meters per second, find the approximate time the rocket would drop to sea level (a height of 0).
Solution:

The equation for this problem is: $y = 250 + 4800x - 4.9x^2$. If the height at sea level is zero, then $y = 0$ so $0 = 250 + 4800x - 4.9x^2$. Solving this for x could be done by using the quadratic formula. In addition, the approximate time in x seconds until the rocket would be at sea level could be estimated by looking at the graph. When the y value of the graph goes from positive to negative then there is a root (also called solution or x intercept) in that interval.

$$x = \frac{^-4800 \pm \sqrt{4800^2 - 4(^-4.9)(250)}}{2(^-4.9)} \approx 980 \text{ or } ^-0.05 \text{ seconds}$$

Since the time has to be positive, it will be about 980 seconds until the rocket is at sea level.

To graph an inequality, graph the quadratic as if it was an equation; however, if the inequality has just a $>$ or $<$ sign, then make the curve itself dotted. Shade above the curve for $>$ or $\geq$. Shade below the curve for $<$ or $\leq$.

Examples:

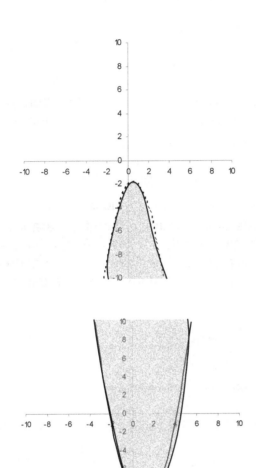

To solve a quadratic equation(with x^2), rewrite the equation into the form:

$$ax^2 + b\,x + c = 0 \quad \text{or} \quad y = ax^2 + b\,x + c$$

where a, b, and c are real numbers. Then substitute the values of a, b, and c into the quadratic formula:

$$x = \frac{-b \pm \sqrt{b^2 - 4ac}}{2a}$$

Simplify the result to find the answers. (Remember, there could be 2 real answers, one real answer, or 2 complex answers that include "i").

To solve a quadratic inequality (with x^2), solve for y. The axis of symmetry is located at $x = {}^-b/2a$. Find coordinates of points to each side of the axis of symmetry. Graph the parabola as a dotted line if the inequality sign is either $<$ or $>$. Graph the parabola as a solid line if the inequality sign is either $\leq$ or $\geq$. Shade above the parabola if the sign is $\geq$ or $>$. Shade below the parabola if the sign is $\leq$ or $<$.

<u>Example</u>: Solve: $8x^2 - 10x - 3 = 0$

In this equation $a = 8$, $b = {}^-10$, and $c = {}^-3$.

Substituting these into the quadratic equation, it becomes:

$$x = \frac{{}^-({}^-10) \pm \sqrt{({}^-10)^2 - 4(8)({}^-3)}}{2(8)} = \frac{10 \pm \sqrt{100 + 96}}{16}$$

$$x = \frac{10 \pm \sqrt{196}}{16} = \frac{10 \pm 14}{16} = 24/16 = 3/2 \text{ or } {}^-4/16 = {}^-1/4$$

<u>Check:</u>

$$x = -\frac{1}{4}$$

$$\frac{1}{2} + \frac{10}{4} - 3 = 0 \qquad \text{Both Check}$$

$$3 - 3 = 0$$

Example: Solve and graph : $y > x^2 + 4x - 5$.

The axis of symmetry is located at $x = {}^-b/2a$. Substituting 4 for b, and 1 for a, this formula becomes:

$$x = {}^-(4)/2(1) = {}^-4/2 = {}^-2$$

Find coordinates of points to each side of $x = {}^-2$.

x	y
$^-5$	0
$^-4$	$^-5$
$^-3$	$^-8$
$^-2$	$^-9$
$^-1$	$^-8$
0	$^-5$
1	0

Graph these points to form a parabola. Draw it as a dotted line. Since a greater than sign is used, shade above and inside the parabola.

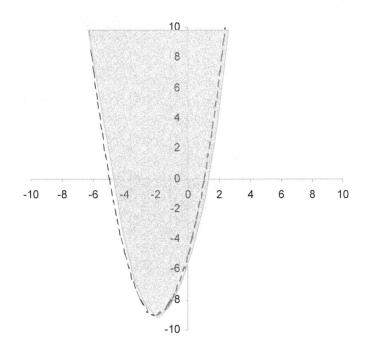

Systems of quadratic equations can be solved by graphing but this method is sometimes inconclusive depending on the accuracy of the graphs and can also be cumbersome. It is recommended that either a substitution or elimination method be considered.

Sample problems:

Find the solution to the system of equations.

1. $y^2 - x^2 = {}^-9$
$2y = x - 3$

1. Use substitution method solving the second equation for x.

$2y = x - 3$
$x = 2y + 3$

2. Substitute this into the first equation in place of (x).

$y^2 - (2y + 3)^2 = {}^-9$

3. Solve.

$y^2 - (4y^2 + 12y + 9) = {}^-9$

$y^2 - 4y^2 - 12y - 9 = {}^-9$

${}^-3y^2 - 12y - 9 = {}^-9$

${}^-3y^2 - 12y = 0$

4. Factor.

${}^-3y(y + 4) = 0$

${}^-3y = 0 \qquad y + 4 = 0$

5. Set each factor equal to zero.

$y = 0 \qquad y = {}^-4$

6. Use these values for y to solve for x.

$2y = x - 3 \qquad 2y = x - 3$

7. Choose an equation.

$2(0) = x - 3 \qquad 2({}^-4) = x - 3$

8. Substitute.

$0 = x - 3 \qquad {}^-8 = x - 3$

$x = 3 \qquad x = {}^-5$

9. Write ordered pairs.

$(3, 0)$ and $({}^-5, {}^-4)$ satisfy the system of equations given.

2. $\begin{array}{l} {}^{-}9x^2 + y^2 = 16 \\ 5x^2 + y^2 = 30 \end{array}$

Use elimination to solve.

$\begin{array}{l} {}^{-}9x^2 + y^2 = 16 \\ \underline{{}^{-}5x^2 - y^2 = {}^{-}30} \\ {}^{-}14x^2 = {}^{-}14 \\ \quad x^2 = 1 \\ \quad x = \pm 1 \end{array}$

1. Multiply second row by

   $^{-}1$.
2. Add.
3. Divide by $^{-}14$.
4. Take the square root of both sides

$\begin{array}{ll} {}^{-}9(1)^2 + y^2 = 16 & {}^{-}9({}^{-}1)^2 + y^2 = 16 \\ {}^{-}9 + y^2 = 16 & {}^{-}9 + y^2 = 16 \\ \quad y^2 = 25 & \quad y^2 = 25 \\ \quad y = \pm 5 & \quad y = \pm 5 \\ \quad (1, \pm 5) & \quad ({}^{-}1, \pm 5) \end{array}$

5. Substitute both values of x into the equation.
6. Take the square root of both sides.
7. Write the ordered pairs.

0008. Understand the properties rational, radical, and absolute value functions and relations.

A function can be defined as a set of ordered pairs in which each element of the domain is paired with one and only one element of the range. The symbol $f(x)$ is read "f of x." Letter other than "f" can be used to represent a function. The letter "g" is commonly used as in $g(x)$.

Sample problems:

1. Given $f(x) = 4x^2 - 2x + 3$, find $f(^-3)$.

(This question is asking for the range value that corresponds to the domain value of $^-3$).

$$f(x) = 4x^2 - 2x + 3$$
$$f(^-3) = 4(^-3)^2 - 2(^-3) + 3$$ 1. Replace x with $^-3$.
$$f(^-3) = 45$$ 2. Solve.

2. Find f(3) and f(10), given $f(x) = 7$.

$$f(x) = 7$$ 1. There are no x values
$$(3) = 7$$ to substitute for. This
     is your answer.

$$f(x) = 7$$
$$f10) = 7$$ 2. Same as above.

Notice that both answers are equal to the constant given.

If $f(x)$ is a function and the value of 3 is in the domain, the corresponding element in the range would be f(3). It is found by evaluating the function for $x = 3$. The same holds true for adding, subtracting, and multiplying in function form.

The symbol f^{-1} is read "the inverse of f". The $^{-1}$ is not an exponent. The inverse of a function can be found by reversing the order of coordinates in each ordered pair that satisfies the function. Finding the inverse functions means switching the place of x and y and then solving for y.

Sample problem:

1. Find $p(a+1) + 3\{p(4a)\}$ if $p(x) = 2x^2 + x + 1$.

Find $p(a+1)$.

$$p(a+1) = 2(a+1)^2 + (a+1) + 1 \qquad \text{Substitute } (a+1) \text{ for } x.$$
$$p(a+1) = 2a^2 + 5a + 4 \qquad \text{Solve.}$$

Find $3\{p(4a)\}$.

$$3\{p(4a)\} = 3[2(4a)^2 + (4a) + 1] \qquad \text{Substitute } (4a) \text{ for } x,$$
$$\text{multiply by 3.}$$
$$3\{p(4a)\} = 96a^2 + 12a + 3 \qquad \text{Solve.}$$

$$p(a+1) + 3\{p(4a)\} = 2a^2 + 5a + 4 + 96a^2 + 12a + 3$$

Combine
like terms.

$$p(a+1) + 3\{p(4a)\} = 98a^2 + 17a + 7$$

The **absolute value function** for a 1st degree equation is of the form:

$y = m(x - h) + k$. Its graph is in the shape of a $\vee$. The point (h,k) is the location of the maximum/minimum point on the graph. "$\pm m$" are the slopes of the 2 sides of the $\vee$. The graph opens up if m is positive and down if m is negative.

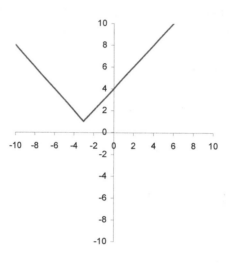

$$y = |x + 3| + 1$$

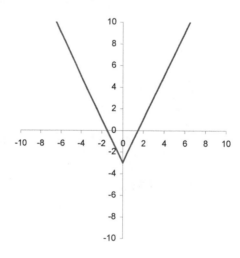

$$y = 2|x| - 3$$

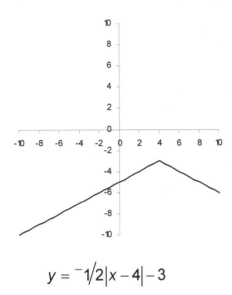

$$y = {}^-1/2|x - 4| - 3$$

-Note that on the first graph above, the graph opens up since m is positive 1. It has $({}^-3,1)$ as its minimum point. The slopes of the 2 upward rays are ± 1.

-The second graph also opens up since m is positive. $(0,{}^-3)$ is its minimum point. The slopes of the 2 upward rays are ± 2.

-The third graph is a downward $\wedge$ because m is ${}^-1/2$. The maximum point on the graph is at $(4,{}^-3)$. The slopes of the 2 downward rays are $\pm 1/2$.

-The **identity function** is the linear equation $y = x$. Its graph is a line going through the origin (0,0) and through the first and third quadrants at a $45°$ degree angle.

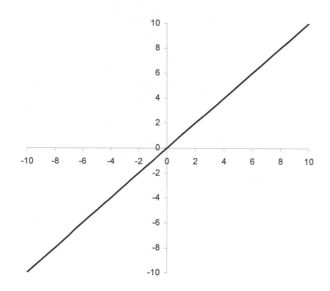

-The **greatest integer function** or **step function** has the equation: $f(x) = j[rx - h] + k$ or $y = j[rx - h] + k$. (h,k) is the location of the left endpoint of one step. j is the vertical jump from step to step. r is the reciprocal of the length of each step. If (x, y) is a point of the function, then when x is an integer, its y value is the same integer. If (x, y) is a point of the function, then when x is not an integer, its y value is the first integer less than x. Points on $y = [x]$ would include:

(3,3), (⁻2,⁻2), (0,0), (1.5,1), (2.83,2), (⁻3.2,⁻4), (⁻.4,⁻1).

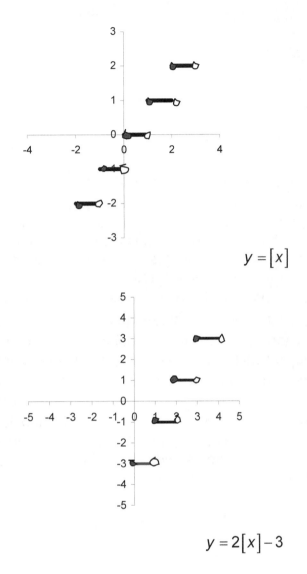

$$y = [x]$$

$$y = 2[x] - 3$$

-Note that in the graph of the first equation, the steps are going up as they move to the right. Each step is one space wide (inverse of r) with a solid dot on the left and a hollow dot on the right where the jump to the next step occurs. Each step is one square higher (j = 1) than the previous step. One step of the graph starts at (0,0) ← values of (h,k).

-In the second graph, the graph goes up to the right. One step starts at the point (0, ⁻3) ← values of (h,k). Each step is one square wide (r = 1) and each step is 2 squares higher than the previous step (j = 2).

Practice: Graph the following equations:

1. $f(x) = x$
2. $y = {}^-|x - 3| + 5$
3. $y = 3[x]$
4. $y = 2/5|x - 5| - 2$

A **rational function** is given in the form $f(x) = p(x)/q(x)$. In the equation, $p(x)$ and $q(x)$ both represent polynomial functions where $q(x)$ does not equal zero. The branches of rational functions approach asymptotes. Setting the denominator equal to zero and solving will give the value(s) of the vertical asymptotes(s) since the function will be undefined at this point. If the value of $f(x)$ approaches b as the $|x|$ increases, the equation $y = b$ is a horizontal asymptote. To find the horizontal asymptote it is necessary to make a table of value for x that are to the right and left of the vertical asymptotes. The pattern for the horizontal asymptotes will become apparent as the $|x|$ increases.

If there are more than one vertical asymptotes, remember to choose numbers to the right and left of each one in order to find the horizontal asymptotes and have sufficient points to graph the function.

Sample problem:

1. Graph $f(x) = \dfrac{3x + 1}{x - 2}$.

$$x - 2 = 0$$
$$x = 2$$

1. Set denominator $= 0$ to find the vertical asymptote.

x	f(x)
3	10
10	3.875
100	3.07
1000	3.007
1	⁻4
⁻10	2.417
⁻100	2.93
⁻1000	2.99

2. Make table choosing numbers to the right and left of the vertical asymptote.

3. Tithe pattern shows that as the $|x|$ increases f(x) approaches the value 3, therefore a horizontal asymptote exists at $y = 3$

Sketch the graph.

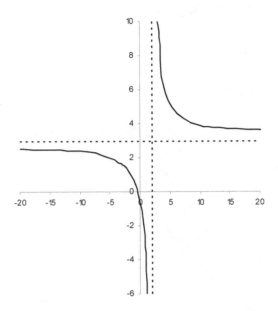

0009. Understand the properties of exponential and logarithmic functions.

When changing common logarithms to exponential form,

$$y = \log_b x \quad \text{if and only if } x = b^y$$

Natural logarithms can be changed to exponential form by using,

$$\log_e x = \ln x \quad \text{or } \ln x = y \quad \text{can be written as } e^y = x$$

Practice Problems:

Express in exponential form.

1. $\log_3 81 = 4$
    $x = 81 \quad b = 3 \quad y = 4$ Identify values.
    $81 = 3^4$ Rewrite in exponential form.

Solve by writing in exponential form.

2. $\log_x 125 = 3$

    $x^3 = 125$ Write in exponential form.
    $x^3 = 5^3$ Write 125 in exponential form.
    $x = 5$ Bases must be equal if exponents
    are equal.

Use a scientific calculator to solve.

3. Find $\ln 72$.
    $\ln 72 = 4.2767$ Use the $\ln x$ key to find natural
    logs.

4. Find $\ln x = 4.2767$ Write in exponential form.
    $e^{4.2767} = x$ Use the key (or 2nd $\ln x$) to find
    x.
    $x = 72.002439$ The small difference is due to
    rounding.

To solve logarithms or exponential functions it is necessary to use several properties.

Multiplication Property $\qquad \log_b mn = \log_b m + \log_b n$

Quotient Property $\qquad \log_b \dfrac{m}{n} = \log_b m - \log_b n$

Powers Property $\qquad \log_b n^r = r \log_b n$

Equality Property $\qquad \log_b n = \log_b m \qquad$ if and only if $n = m$.

Change of Base Formula $\qquad \log_b n = \dfrac{\log n}{\log b}$

$$\log_b b^x = x \text{ and } b^{\log_b x} = x$$

Sample problem.

Solve for x.
1. $\log_6(x-5)+\log_6 x = 2$

    $\log_6 x(x-5) = 2$ Use product property.

    $\log_6 x^2 - 5x = 2$ Distribute.

    $x^2 - 5x = 6^2$ Write in exponential form.

    $x^2 - 5x - 36 = 0$ Solve quadratic equation.

    $(x+4)(x-9) = 0$

      $x = {}^-4 \quad x = 9$

***Be sure to check results. Remember x must be greater than zero in $\log x = y$.

Check: $\log_6(x-5)+\log_6 x = 2$

        $\log_6({}^-4-5)+\log_6({}^-4) = 2$ Substitute the first answer ${}^-4$.

        $\log_6({}^-9)+\log_6({}^-4) = 2$ This is undefined, x is less than zero.

        $\log_6(9-5)+\log_6 9 = 2$ Substitute the second answer 9.

        $\log_6 4 + \log_6 9 = 2$

        $\log_6(4)(9) = 2$ Multiplication property.

        $\log_6 36 = 2$

        $6^2 = 36$ Write in exponential form.

        $36 = 36$

Practice problems:

2. $\log_4 x = 2\log_4 3$

3. $2\log_3 x = 2 + \log_3(x-2)$

4. Use change of base formula to find $(\log_3 4)(\log_4 3)$.

Coefficients are the numbers in front of the variables in a term.

Addition/Subtraction with integral exponents.
-**Like terms** have the same variables raised to the same powers. Like terms can be combined by adding or subtracting their coefficients and by keeping the variables and their exponents the same. Unlike terms cannot be combined.

$$5x^4 + 3x^4 - 2x^4 = 6x^4$$
$$3ab^2 + 5ab + ab^2 = 4ab^2 + 5ab$$

Multiplication with integral exponents.
-When multiplying terms together, multiply their coefficients and add the exponents of the same variables. When a term is raised to a power, raise the term's coefficient to the power outside the parentheses. Then multiply the outside exponent times each of the inside exponents.

$$(3a^7b^5)(^-4a^2b^9c^4) = {}^-12a^9b^{14}c^4$$
$$(^-2x^3yz^7)^5 = {}^-32x^{15}y^5z^{35}$$

Division with integral exponents.
-When dividing any number of terms by a single term, divide or reduce their coefficients. Then subtract the exponent of a variable on the bottom from the exponent of the same variable on the top.

$$\frac{24a^8b^7 + 16a^7b^5 - 8a^6}{8a^6} = 3a^2b^7 + 2ab^5 - 1$$

Negative exponents are usually changed into positive exponents by moving those variables from numerator to denominator (or vice versa) to make the exponents become positive.

$$\frac{^-60a^8b^{-7}c^{-6}d^5e^{-12}}{12a^{-2}b^3c^{-9}d^5e^{-4}} = {}^- 5a^{10}b^{-10}c^3e^{-8} = \frac{^-5a^{10}c^3}{b^{10}e^8}$$

$$(6x^{-9}y^3)(^-2x^5y^{-1}) = {}^- 12x^{-4}y^2 = \frac{^-12y^2}{x^4}$$

Simplify as far as possible:
1. $(4x^4y^2)(^-3xy^{-3}z^5)$
2. $(100a^6 + 60a^4 - 28a^2b^3) \div (4a^2)$
3. $(3x^5y^2)^4 + 8x^9y^6 + 19x^{20}y^8$

SUBAREA III–TRIGONOMETRY AND CALCULUS

0010. Understand principles, properties, and relationships involving trigonometric functions and their associated geometric representations

It is easiest to graph trigonometric functions when using a calculator by making a table of values.

DEGREES

	0	30	45	60	90	120	135	150	180	210	225	240	270	300	315	330	360
sin	0	.5	.71	.87	1	.87	.71	.5	0	-.5	-.71	-.87	-1	-.87	-.71	-.5	0
cos	1	.87	.71	.5	0	-.5	-.71	-.87	-1	-.87	-.71	-.5	0	.5	.71	.87	1
tan	0	.58	1	1.7	--	-1.7	-1	-.58	0	.58	1	1.7	--	-1.7	-1	-.58	0

$$0 \quad \frac{\pi}{6} \quad \frac{\pi}{4} \quad \frac{\pi}{3} \quad \frac{\pi}{2} \quad \frac{2\pi}{3} \quad \frac{3\pi}{4} \quad \frac{5\pi}{6} \quad \pi \quad \frac{7\pi}{6} \quad \frac{5\pi}{4} \quad \frac{4\pi}{3} \quad \frac{3\pi}{2} \quad \frac{5\pi}{3} \quad \frac{7\pi}{4} \quad \frac{11\pi}{6} \quad 2\pi$$

RADIANS

Remember the graph always ranges from +1 to ⁻1 for sine and cosine functions unless noted as the coefficient of the function in the equation. For example, $y = 3\cos x$ has an amplitude of 3 units from the center line (0). Its maximum and minimum points would be at +3 and ⁻3.

Tangent is not defined at the values 90 and 270 degrees or $\frac{\pi}{2}$ and $\frac{3\pi}{2}$. Therefore, vertical asymptotes are drawn at those values.

The inverse functions can be graphed in the same manner using a calculator to create a table of values.

Given the special right triangles below, we can find the lengths of other special right triangles.

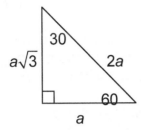 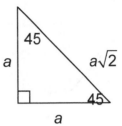

Sample problems:

1. 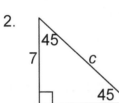   if $8 = a\sqrt{2}$ then $a = 8/\sqrt{2}$ or 5.657

2.   if $7 = a$ then $c = a\sqrt{2} = 7\sqrt{2}$ or 9.899

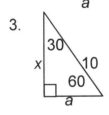

3.   if $2a = 10$ then $a = 5$ and $x = a\sqrt{3} = 5\sqrt{3}$ or 8.66

Given right triangle ABC, the adjacent side and opposite side can be identified for each angle A and B.

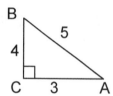

Looking at angle A, it can be determined that side *b* is adjacent to angle A and side *a* is opposite angle A.

If we now look at angle B, we see that side *a* is adjacent to angle *b* and side *b* is opposite angle B.

The longest side (opposite the 90 degree angle) is always called the hypotenuse.

The basic trigonometric ratios are listed below:

$$\text{Sine} = \frac{\text{opposite}}{\text{hypotenuse}}, \text{Cosine} = \frac{\text{adjacent}}{\text{hypotenuse}}, \text{Tangent} = \frac{\text{opposite}}{\text{adjacent}}$$

Sample problem:

1. Use triangle ABC to find the sin, cos and tan for angle A.

sin A= 4/5
cos A = 3/5
tan A = 4/3

Use the basic trigonometric ratios of sine, cosine and tangent to solve for the missing sides of right triangles when given at least one of the acute angles.

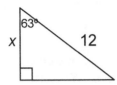

In the triangle ABC, an acute angle of 63 degrees and the length of the hypotenuse (12). The missing side is the one adjacent to the given angle.

The appropriate trigonometric ratio to use would be cosine since we are looking for the adjacent side and we have the length of the hypotenuse.

$$Cos x = \frac{adjacent}{hypotenuse}$$
  1. Write formula.

$$Cos 63 = \frac{x}{12}$$
  2. Substitute known values.

$$0.454 = \frac{x}{12}$$

$$x = 5.448$$
  3. Solve.

Unlike trigonometric identities that are true for all values of the defined variable, trigonometric equations are true for some, but not all, of the values of the variable. Most often trigonometric equations are solved for values between 0 and 360 degrees or 0 and 2π radians.

Some algebraic operation, such as squaring both sides of an equation, will give you extraneous answers. You must remember to check all solutions to be sure that they work.

Sample problems:

1. Solve: $\cos x = 1 - \sin x$ if $0 \leq x < 360$ degrees.

$\cos^2 x = (1 - \sin x)^2$	1. square both sides
$1 - \sin^2 x = 1 - 2\sin x + \sin^2 x$	2. substitute
$0 = {}^-2\sin x + 2\sin^2 x$	3. set $=$ to 0
$0 = 2\sin x({}^-1 + \sin x)$	4. factor
$2\sin x = 0 \qquad {}^-1 + \sin x = 0$	5. set each factor $= 0$
$\sin x = 0 \qquad\qquad \sin x = 1$	6. solve for $\sin x$
$x = 0 \text{ or } 180 \qquad x = 90$	7. find value of sin at x

The solutions appear to be 0, 90 and 180. Remember to check each solution and you will find that 180 does not give you a true equation. Therefore, the only solutions are 0 and 90 degrees.

2. Solve: $\cos^2 x = \sin^2 x$ if $0 \leq x < 2\pi$

$\cos^2 x = 1 - \cos^2 x$	1. substitute
$2\cos^2 x = 1$	2. simplify
$\cos^2 x = \dfrac{1}{2}$	3. divide by 2
$\sqrt{\cos^2 x} = \pm\sqrt{\dfrac{1}{2}}$	4. take square root
$\cos x = \dfrac{\pm\sqrt{2}}{2}$	5. rationalize denominator

$$x = \frac{\pi}{4}, \frac{3\pi}{4}, \frac{5\pi}{4}, \frac{7\pi}{4}$$

Given 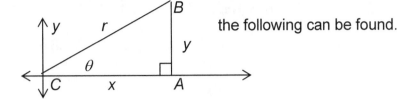 the following can be found.

Trigonometric Functions:

$$\sin\theta = \frac{y}{r} \qquad\qquad \csc\theta = \frac{r}{y}$$

$$\cos\theta = \frac{x}{r} \qquad\qquad \sec\theta = \frac{r}{x}$$

$$\tan\theta = \frac{y}{x} \qquad\qquad \cot\theta = \frac{x}{y}$$

Sample problem:

1. Prove that $\sec\theta = \dfrac{1}{\cos\theta}$.

$$\sec\theta = \frac{1}{\frac{x}{r}} \qquad\qquad\text{Substitution definition of cosine.}$$

$$\sec\theta = \frac{1\times r}{\frac{x}{r}\times r} \qquad\qquad\text{Multiply by } \frac{r}{r}.$$

$$\sec\theta = \frac{r}{x} \qquad\qquad\text{Substitution.}$$

$$\sec\theta = \sec\theta \qquad\qquad\text{Substitute definition of } \frac{r}{x}.$$

$$\sec\theta = \frac{1}{\cos\theta} \qquad\qquad\text{Substitute.}$$

2. Prove that $\sin^2 + \cos^2 = 1$.

$$\left(\frac{y}{r}\right)^2 + \left(\frac{x}{r}\right)^2 = 1 \qquad$$ Substitute definitions of sin and cos.

$$\frac{y^2 + x^2}{r^2} = 1 \qquad x^2 + y^2 = r^2 \text{ Pythagorean}$$

formula.

$$\frac{r^2}{r^2} = 1 \qquad$$ Simplify.

$$1 = 1 \qquad$$ Substitute.

$$\sin^2\theta + \cos^2\theta = 1$$

Practice problems: Prove each identity.

1. $\cot\theta = \dfrac{\cos\theta}{\sin\theta}$ 　　　　　2. $1 + \cot^2\theta = \csc^2\theta$

There are two methods that may be used to prove **trigonometric identities**. One method is to choose one side of the equation and manipulate it until it equals the other side. The other method is to replace expressions on both sides of the equation with equivalent expressions until both sides are equal.

The Reciprocal Identities

$$\sin x = \frac{1}{\csc x} \qquad\qquad \sin x\, \csc x = 1 \qquad\qquad \csc x = \frac{1}{\sin x}$$

$$\cos x = \frac{1}{\sec x} \qquad\qquad \cos x\, \sec x = 1 \qquad\qquad \sec x = \frac{1}{\cos x}$$

$$\tan x = \frac{1}{\cot x} \qquad\qquad \tan x\, \cot x = 1 \qquad\qquad \cot x = \frac{1}{\tan x}$$

$$\tan x = \frac{\sin x}{\cos x} \qquad\qquad\qquad\qquad\qquad\qquad \cot x = \frac{\cos x}{\sin x}$$

The Pythagorean Identities

$$\sin^2 x + \cos^2 x = 1 \qquad 1 + \tan^2 x = \sec^2 x \qquad 1 + \cot^2 x = \csc^2 x$$

Sample problems:

1. Prove that $\cot x + \tan x = (\csc x)(\sec x)$.

$$\frac{\cos x}{\sin x} + \frac{\sin x}{\cos x}$$

Reciprocal identities.

$$\frac{\cos^2 x + \sin^2 x}{\sin x \cos x}$$

Common denominator.

$$\frac{1}{\sin x \cos x}$$

Pythagorean identity.

$$\frac{1}{\sin x} \times \frac{1}{\cos x}$$
$$\csc x(\sec x) = \csc x(\sec x)$$

Reciprocal identity, therefore,

$$\cot x + \tan x = \csc x(\sec x)$$

2. Prove that $\dfrac{\cos^2 \theta}{1 + 2\sin\theta + \sin^2 \theta} = \dfrac{\sec\theta - \tan\theta}{\sec\theta + \tan\theta}$.

$$\frac{1 - \sin^2 \theta}{(1 + \sin\theta)(1 + \sin\theta)} = \frac{\sec\theta - \tan\theta}{\sec\theta + \tan\theta}$$

Pythagorean identity factor denominator.

$$\frac{1 - \sin^2 \theta}{(1 + \sin\theta)(1 + \sin\theta)} = \frac{\dfrac{1}{\cos\theta} - \dfrac{\sin\theta}{\cos\theta}}{\dfrac{1}{\cos\theta} + \dfrac{\sin\theta}{\cos\theta}}$$

Reciprocal

identities.

$$\frac{(1 - \sin\theta)(1 + \sin\theta)}{(1 + \sin\theta)(1 + \sin\theta)} = \frac{\dfrac{1 - \sin\theta}{\cos\theta}(\cos\theta)}{\dfrac{1 + \sin\theta}{\cos\theta}(\cos\theta)}$$

Factor $1 - \sin^2 \theta$.

Multiply by $\dfrac{\cos\theta}{\cos\theta}$.

$$\frac{1 - \sin\theta}{1 + \sin\theta} = \frac{1 - \sin\theta}{1 + \sin\theta}$$

Simplify.

$$\frac{\cos^2 \theta}{1 + 2\sin\theta + \sin^2 \theta} = \frac{\sec\theta - \tan\theta}{\sec\theta + \tan\theta}$$

Trigonometric functions can be expanded in **power series**, which facilitates approximations of the functions in extreme cases, such as situations involving very small angles. The angle x must be in radians. The formula for a power series is:

$$\sin x = x - \frac{x^3}{3!} + \frac{x^5}{5!} - \frac{x^7}{7!} + \dots \quad \text{For small } x,\ \sin x \approx x.$$

$$\cos x = 1 - \frac{x^2}{2!} + \frac{x^4}{4!} - \frac{x^6}{6!} + \dots \quad \text{For small } x,\ \cos x \approx 1.$$

$$\tan x = x + \frac{x^3}{3} + \frac{2x^5}{15} + \frac{17x^7}{315} + \dots \quad \text{For small } x,\ \tan x \approx x.$$

In situations with very small angles, the trigonometric functions can be approximated by the first term in their series.

The basic **exponential function** is defined by $f(x) = B^x$. Stated another way, it is the function consisting of the base of the natural logarithm e taken to the power of the variable. e is the constant 2.718....

The exponential function satisfies the identities,

$$e^x = \cosh x + \sinh x$$
$$e^x = \sec(\text{gd}x) + \tan(\text{gd}x)$$
$$e^x = \tan\left(\frac{1}{4}\pi + \frac{1}{2}\text{gd}x\right)$$
$$e^x = \frac{1 + \sin(\text{gd}x)}{\cos(\text{gd}x)}$$

where $\text{gd}x$ is the **Gudermannian function**. The Gudermannian function is defined:

$$\text{gd}x = 2\tan^{-1}\left(e^x\right) - \frac{\pi}{2}.$$

The **complex exponential function** is defined as
$e^{i\theta} = \cos\theta + i\sin\theta$.

Example: Explore the properties of trigonometric functions using complex exponential functions.

You know that:

$e^{i(a+b)} = e^{ia}e^{ib}$.

Using exponential/trigonometric equivalence, we get

$e^{i(a+b)} = \cos(a+b) + i\sin(a+b)$

and

$e^{ia}e^{ib} = (\cos(a) + i\sin(a))(\cos(b) + i\sin(b))$
$= \cos(a)\cos(b) - \sin(a)\sin(b) + i(\sin(a)\cos(b) + \cos(a)\sin(b))$.

Comparing these two, we get two separate trigonometric identities

$\cos(a+b) = \cos(a)\cos(b) - \sin(a)\sin(b)$
$\sin(a+b) = \sin(a)\cos(b) + \cos(a)\sin(b)$.

The trigonometric functions sine, cosine, and tangent are **periodic functions**. The values of periodic functions repeat on regular intervals. Period, amplitude, and phase shift are key properties of periodic functions that can be determined by observation of the graph.

The **period** of a function is the smallest domain containing the complete cycle of the function. For example, the period of a sine or cosine function is the distance between the peaks of the graph.

The **amplitude** of a function is half the distance between the maximum and minimum values of the function.

Phase shift is the amount of horizontal displacement of a function from its original position.

Below is a generic sine/cosine graph with the period and amplitude labeled.

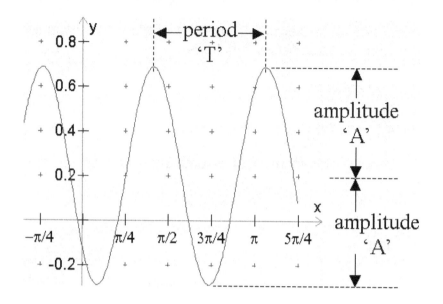

Properties of the graphs of basic trigonometric functions.

Function	Period	Amplitude
$y = \sin x$	2π radians	1
$y = \cos x$	2π radians	1
$y = \tan x$	π radians	undefined

Below are the graphs of the basic trigonometric functions, (a) $y = \sin x$; (b) $y = \cos x$; and (c) $y = \tan x$.

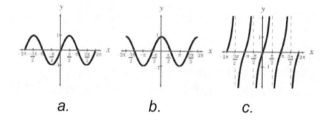

| a. | b. | c. |

Note that the phase shift of trigonometric graphs is the horizontal distance displacement of the curve from these basic functions.

0011. **Understand the principles and techniques of trigonometry to model and solve problems**

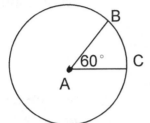

Central angle BAC = 60°
Minor arc BC = 60°
Major arc BC = 360 − 60 = 300°

If you draw two radii in a circle, the angle they form with the center as the vertex is a central angle. The piece of the circle "inside" the angle is an arc. Just like a central angle, an arc can have any degree measure from 0 to 360. The measure of an arc is equal to the measure of the central angle which forms the arc. Since a diameter forms a semicircle and the measure of a straight angle like a diameter is 180°, the measure of a semicircle is also 180°.

Given two points on a circle, there are two different arcs which the two points form. Except in the case of semicircles, one of the two arcs will always be greater than 180° and the other will be less than 180°. The arc less than 180° is a minor arc and the arc greater than 180° is a major arc.

Examples:

1.

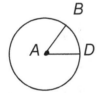

$m\angle BAD = 45°$
What is the measure of the major arc BD?

$\angle BAD$ = minor arc BD

45° = minor arc BD

360 − 45 = major arc BD

315° = major arc BD

The measure of the central angle is the same as the measure of the arc it forms. A major and minor arc always add to 360°.

2.

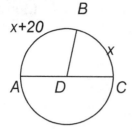

$\overline{AC}$ is a diameter of circle D.
What is the measure of $\measuredangle BDC$?

$$m\measuredangle ADB + m\measuredangle BDC = 180°$$
$$x + 20 + x = 180$$
$$2x + 20 = 180$$
$$2x = 160$$
$$x = 80$$

A diameter forms a semicircle
which has a measure of $180°$.

minor arc $BC = 80°$
$$m\measuredangle BDC = 80°$$

A central angle has the same
measure as the arc it forms.

$$\frac{\measuredangle PQR}{360°} = \frac{\text{length of arc } RP}{\text{circumference of } \bigcirc Q} = \frac{\text{area of sector } PQR}{\text{area of } \bigcirc Q}$$

While an arc has a measure associated to the degree measure of a central angle, it also has a length which is a fraction of the circumference of the circle.

For each central angle and its associated arc, there is a sector of the circle which resembles a pie piece. The area of such a sector is a fraction of the area of the circle.

The fractions used for the area of a sector and length of its associated arc are both equal to the ratio of the central angle to 360°.

Examples:

1.

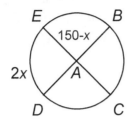

$\bigcirc A$ has a radius of 4 cm. What is the length of arc ED?

$$2x + 150 - x = 180$$
$$x + 150 = 180$$
$$x = 30$$

Arc BE and arc DE make a semicircle.

Arc $ED = 2(30) = 60°$

The ratio $60°$ to $360°$ is equal to the ratio of arch length ED to the circumference of $\bigcirc A$.

$$\frac{60}{360} = \frac{\text{arc length } ED}{2\pi 4}$$

$$\frac{1}{6} = \frac{\text{arc length}}{8\pi}$$

Cross multiply and solve for the arc length.

$$\frac{8\pi}{6} = \text{arc length}$$

$$\text{arc length } ED = \frac{4\pi}{3} \text{ cm}.$$

2.

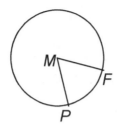

The radius of $\bigcirc M$ is 3 cm. The length of arc PF is 2π cm. What is the area of sector PMF?

Circumference of $\bigcirc M = 2\pi(3) = 6\pi$

Area of $\bigcirc M = \pi(3)^2 = 9\pi$

Find the circumference and area of the circle.

$$\frac{\text{area of } MPF}{9\pi} = \frac{2\pi}{6\pi}$$

The ratio of the sector area to the circle area is the same as the arc length to the circumference.

$$\frac{\text{area of } MPF}{9\pi} = \frac{1}{3}$$

$$\text{area of } MPF = \frac{9\pi}{3}$$

$$\text{area of } MPF = 3\pi$$

Solve for the area of the sector.

A tangent line intersects a circle in exactly one point. If a radius is drawn to that point, the radius will be perpendicular to the tangent.

A **chord** is a segment with endpoints on the circle. If a radius or diameter is perpendicular to a chord, the radius will cut the chord into two equal parts.

If **two chords** in the same circle have the same length, the two chords will have arcs that are the same length, and the two chords will be equidistant from the center of the circle. Distance from the center to a chord is measured by finding the length of a segment from the center perpendicular to the chord.

Examples:

1.

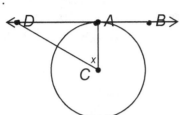

$\overrightarrow{DB}$ is tangent to $\angle C$ at A.
$m\angle ADC = 40°$ Find x.

$\overline{AC} \perp \overrightarrow{DB}$ — A radius is $\perp$ to a tangent at the point of tangency.

$m\angle DAC = 90°$ — Two segments that are $\perp$ form a $90°$ angle.

$40 + 90 + x = 180$ — The sum of the angles of a triangle is $180°$.

$x = 50°$ — Solve for x.

2.

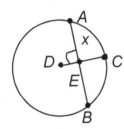

$\overline{CD}$ is a radius and $\overline{CD} \perp$ chord $\overline{AB}$.
$\overline{AB} = 10$. Find x.

$x = \dfrac{1}{2}(10)$

$x = 5$ — If a radius is $\perp$ to a chord, the radius bisects the chord.

$X = 5$

Angles with their vertices in a circle's interior:

When two chords intersect inside a circle, two sets of vertical angles are formed. Each set of vertical angles intercepts two arcs which are across from each other. The measure of an angle formed by two chords in a circle is equal to one-half the sum of the angle intercepted by the angle and the arc intercepted by its vertical angle.

Angles with their vertices in a circle's exterior:

If an angle has its vertex outside of the circle and each side of the circle intersects the circle, then the angle contains two different arcs. The measure of the angle is equal to one-half the difference of the two arcs.

Examples:
1.

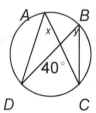

Find x *and* y.

$$m\angle DAC = \frac{1}{2}(40) = 20°$$

$\angle DAC$ and $\angle DBC$ are both inscribed angles, so each one has a measure equal to one-half the measure of arc *DC*.

$$m\angle DBC = \frac{1}{2}(40) = 20°$$

$$x = 20° \text{ and } y = 20°$$

Intersecting chords:

If two chords intersect inside a circle, each chord is divided into two smaller segments. The product of the lengths of the two segments formed from one chord equals the product of the lengths of the two segments formed from the other chord.

Intersecting tangent segments:

If two tangent segments intersect outside of a circle, the two segments have the same length.

Intersecting secant segments:

If two secant segments intersect outside a circle, a portion of each segment will lie inside the circle and a portion (called the exterior segment) will lie outside the circle. The product of the length of one secant segment and the length of its exterior segment equals the product of the length of the other secant segment and the length of its exterior segment.

Tangent segments intersecting secant segments:

If a tangent segment and a secant segment intersect outside a circle, the square of the length of the tangent segment equals the product of the length of the secant segment and its exterior segment.

Examples:

1.

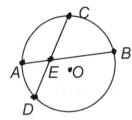

$\overline{AB}$ and $\overline{CD}$ are chords.
$CE=10$, $ED=x$, $AE=5$, $EB=4$

$$(AE)(EB) = (CE)(ED)$$

Since the chords intersect in the circle, the products of the segment pieces are equal.

$$5(4) = 10x$$
$$20 = 10x$$
$$x = 2$$

Solve for x.

2.

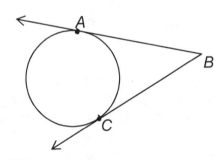

$\overline{AB}$ and $\overline{CD}$ are chords.
$\overline{AB} = x^2 + x - 2$
$\overline{BC} = x^2 - 3x + 5$
Find the length of
$\overline{AB}$ and $\overline{BC}$.

$\overline{AB} = x^2 + x - 2$
$\overline{BC} = x^2 - 3x + 5$

Given

$\overline{AB} = \overline{BC}$

Intersecting tangents are equal.

$x^2 + x - 2 = x^2 - 3x + 5$

Set the expression equal and solve.

$4x = 7$
$x = 1.75$

Substitute and solve.

$(1.75)^2 + 1.75 - 2 = \overline{AB}$
$\overline{AB} = \overline{BC} = 2.81$

Unlike trigonometric identities that are true for all values of the defined variable, trigonometric equations are true for some, but not all, of the values of the variable. Most often trigonometric equations are solved for values between 0 and 360 degrees or 0 and 2π radians.

Some algebraic operation, such as squaring both sides of an equation, will give you extraneous answers. You must remember to check all solutions to be sure that they work.

Sample problems:

1. Solve: $\cos x = 1 - \sin x$ if $0 \le x < 360$ degrees.

$\cos^2 x = (1 - \sin x)^2$ 1. square both sides

$1 - \sin^2 x = 1 - 2\sin x + \sin^2 x$ 2. substitute

$0 = {}^{-}2\sin x + 2\sin^2 x$ 3. set = to 0

$0 = 2\sin x({}^{-}1 + \sin x)$ 4. factor

$2\sin x = 0$ $^{-}1 + \sin x = 0$ 5. set each factor = 0

$\sin x = 0$ $\sin x = 1$ 6. solve for $\sin x$

$x = 0$ or 180 $x = 90$ 7. find value of $\sin$ at x

The solutions appear to be 0, 90 and 180. Remember to check each solution and you will find that 180 does not give you a true equation. Therefore, the only solutions are 0 and 90 degrees.

2. Solve: $\cos^2 x = \sin^2 x$ if $0 \le x < 2\pi$

$\cos^2 x = 1 - \cos^2 x$ 1. substitute

$2\cos^2 x = 1$ 2. simplify

$\cos^2 x = \dfrac{1}{2}$ 3. divide by 2

$\sqrt{\cos^2 x} = \pm\sqrt{\dfrac{1}{2}}$ 4. take square root

$\cos x = \dfrac{\pm\sqrt{2}}{2}$ 5. rationalize denominator

$x = \dfrac{\pi}{4}, \dfrac{3\pi}{4}, \dfrac{5\pi}{4}, \dfrac{7\pi}{4}$

In order to solve a right triangle using trigonometric functions it is helpful to identify the given parts and label them. Usually more than one trigonometric function may be appropriately applied.

Some items to know about right triangles:

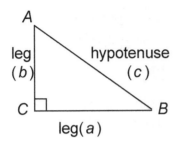

Given angle A, the side labeled leg (b) is adjacent angle A. And the side labeled leg (a) is opposite to angle A.

Sample problem:

1. Find the missing side.

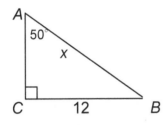

1. Identify the known values. Angle $A = 50$ degrees and the side opposite the given angle is 12. The missing side is the adjacent leg.

2. The information suggests the use of the tangent function

$$\tan A = \frac{\text{opposite}}{\text{adjacent}}$$

3. Write the function.

$$\tan 50 = \frac{12}{x}$$

4. Substitute.

$$1.192 = \frac{12}{x}$$

5. Solve.

$$x(1.192) = 12$$
$$x = 10.069$$

Remember that since angle A and angle B are complimentary, then angle $B = 90 - 50$ or 40 degrees.

Using this information we could have solved for the same side only this time it is the leg opposite from angle B.

$$\tan B = \frac{\text{opposite}}{\text{adjacent}}$$ 1. Write the formula.

$$\tan 40 = \frac{x}{12}$$ 2. Substitute.

$12(.839) = x$ 3. Solve.

$10.069 \approx x$

Now that the two sides of the triangle are known, the third side can be found using the Pythagorean Theorem.

Definition: For any triangle ABC, when given two sides and the included angle, the other side can be found using one of the formulas below:

$$a^2 = b^2 + c^2 - (2bc)\cos A$$
$$b^2 = a^2 + c^2 - (2ac)\cos B$$
$$c^2 = a^2 + b^2 - (2ab)\cos C$$

Similarly, when given three sides of a triangle, the included angles can be found using the derivation:

$$\cos A = \frac{b^2 + c^2 - a^2}{2bc}$$

$$\cos B = \frac{a^2 + c^2 - b^2}{2ac}$$

$$\cos C = \frac{a^2 + b^2 - c^2}{2ab}$$

Sample problem:

1. Solve triangle ABC, if angle $B = 87.5°$, $a = 12.3$, and $c = 23.2$. (Compute to the nearest tenth).

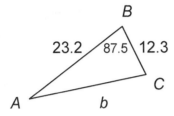

1. Draw and label a sketch.

Find side b.

$$b^2 = a^2 + c^2 - (2ac)\cos B$$

2. Write the formula.

$$b^2 = (12.3)^2 + (23.2)^2 - 2(12.3)(23.2)(\cos 87.5)$$

3. Substitute.

$$b^2 = 664.636$$

$$b = 25.8 \text{ (rounded)}$$

4. Solve.

Use the law of sines to find angle A.

$$\frac{\sin A}{a} = \frac{\sin B}{b}$$

1. Write formula.

$$\frac{\sin A}{12.3} = \frac{\sin 87.5}{25.8} = \frac{12.29}{25.8}$$

2. Substitute.

$$\sin A = 0.47629$$

3. Solve.

Angle $A = 28.4$

Therefore, angle $C = 180 - (87.5 + 28.4)$
$$= 64.1$$

2. Solve triangle ABC if $a = 15$, $b = 21$, and $c = 18$. (Round to the nearest tenth).

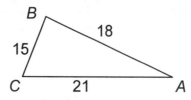

1. Draw and label a sketch.

Find angle A.

$$\cos A = \frac{b^2 + c^2 - a^2}{2bc}$$

2. Write formula.

$$\cos A = \frac{21^2 + 18^2 - 15^2}{2(21)(18)}$$

3. Substitute.

$\cos A = 0.714$

4. Solve.

Angle $A = 44.4$

Find angle B.

$$\cos B = \frac{a^2 + c^2 - b^2}{2ac}$$

5. Write formula.

$$\cos B = \frac{15^2 + 18^2 - 21^2}{2(15)(18)}$$

6. Substitute.

$\cos B = 0.2$

7. Solve.

Angle $B = 78.5$

Therefore, angle $C = 180 - (44.4 + 78.5)$
$$= 57.1$$

Definition: For any triangle ABC, where a, b, and c are the lengths of the sides opposite angles A, B, and C respectively:

$$\frac{\sin A}{a} = \frac{\sin B}{b} = \frac{\sin C}{c}$$

Sample problem:

1. An inlet is 140 feet wide. The lines of sight from each bank to an approaching ship are 79 degrees and 58 degrees. What are the distances from each bank to the ship?

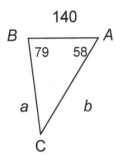

1. Draw and label a sketch.

2. $180 - (79 + 58) = 43$ degrees.

$$\frac{\sin A}{a} = \frac{\sin B}{b} = \frac{\sin C}{c}$$

3. Write formula.

Side opposite 79 degree angle:

$$\frac{\sin 79}{b} = \frac{\sin 43}{140}$$

4. Substitute.

$$b = \frac{140(.9816)}{.6820}$$

5. Solve.

$$b \approx 201.501 \text{ feet}$$

Side opposite 58 degree angle:

$$\frac{\sin 58}{a} = \frac{\sin 43}{140}$$

6. Substitute.

$$a = \frac{140(.848)}{.6820}$$

7. Solve.

$$a \approx 174.076 \text{ feet}$$

When the measure of two sides and an angle not included between the sides are given, there are several possible solutions. There may be one, two or no triangles that may be formed from the given information.

The ambiguous case is described using two situations: either angle A is acute or it is obtuse.

Case 1: Angle A is acute.

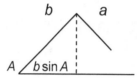

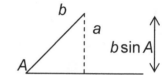

$a < b(\sin A)$
No triangle possible.

$a = b(\sin A)$
One triangle.

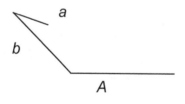

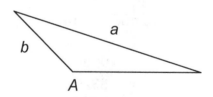

$a > b(\sin A)$ and $a \geq b$
One triangle.

$a > b(\sin A)$ but $a < b$
Two triangles.

Case 2: Angle A is obtuse.

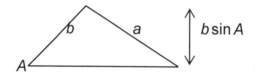

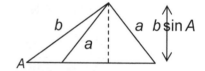

$a \leq b$
No triangle.

$a > b$
One triangle.

Sample problems:

Determine how many solutions exist.

1. Angle $A = 70$ degrees, $a = 16$, $b = 21$.
Angle A is an acute angle and $a < b$ so we are looking at a case 1 triangle. In this case, $b(\sin A) = 19.734$. Since $a = 16$, this gives us a triangle possibility similar to case 1 triangle 1 where $a < b(\sin A)$ and there are no possible solutions.

2. Angle $C = 95.1$ degrees, $b = 16.8$, and $c = 10.9$.
Angle C is an obtuse angle and $c \leq b$ (case 2, triangle 1). There are no possible solutions.

3. Angle $B = 45$ degrees, $b = 40$, and $c = 32$.
Angle B is acute and $b \geq c$. Finding $c(\sin B)$ gives 22.627 and therefore, $b > c(\sin B)$. This indicates a case 1 triangle with one possible solution.

Find the number of possible solutions and then the missing sides, if possible (round all answers to the nearest whole number).

4. Angle $A = 37$ degrees, $a = 49$ and $b = 54$.
Angle A is acute and $a < b$, find $b(\sin A)$. If $a > b(\sin A)$ there will be two triangles, and if $a < b(\sin A)$ there will be no triangles possible.

$a = 49$ and $b(\sin A) = 32.498$, therefore, $a > b(\sin A)$ and there are two triangle solutions.

Use law of sines to find angle B.

$$\frac{\sin A}{a} = \frac{\sin B}{b}$$ 1. Write formula.

$$\frac{\sin 37}{49} = \frac{\sin B}{54}$$ 2. Substitute.

$$\sin B = 0.663$$ 3. Solve.

Angle $B = 42$ degrees or angle $B = 180 - 42 = 138$ degrees. There are two possible solutions to be solved for.

Case 1 (angle $B = 42$ degrees)

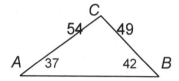

1. Draw sketch

Angle $C = 180 - (37 + 42)$
$\qquad = 101 \quad$ degrees

2. Find angle C.

$$\frac{\sin A}{a} = \frac{\sin C}{c}$$

3. Find side c using law of sines.

$$\frac{\sin 37}{49} = \frac{\sin 101}{c}$$

4. Substitute.

$c = 79.924 \approx 80$

Case 2 (angle $B = 138$)

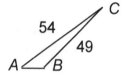

1. Draw a sketch.
   ($A = 37$ and $B = 138$)

Angle $C = 180 - (37 + 138)$
$\qquad = 5 \quad$ degrees

2. Find angle C.

$$\frac{\sin A}{a} = \frac{\sin C}{c}$$

3. Find side c using law of sine.

$$\frac{\sin 37}{49} = \frac{\sin 5}{c}$$

4. Substitute.

$c = 7.097 \approx 7$

When finding the missing sides and angles of triangles that are either acute or obtuse using the law of sines and the law of cosines is imperative. Below is a chart to assist in determining the correct usage.

Given	Suggested Solution Method
(SAS) Two sides and the included angle	Law of Cosines will give you the third side. Then use the Law of Sines for the angles.
(SSS) Three sides	Law of Cosines will give you an angle. Then the Law of Sines can be used for the other angles.
(SAA or ASA) One side, two angles	Find the remaining angle and then use the Law of Sines.
(SSA) Two sides, angle not included	Find the number of possible solutions. Use the Law of Sines.

Definition: The area of any triangle ABC can be found using one of these formulas when given two legs and the included angle:

$$\text{Area} = \frac{1}{2}bc\sin A$$

$$\text{Area} = \frac{1}{2}ac\sin B$$

$$\text{Area} = \frac{1}{2}ab\sin C$$

Sample problem:

Find the area of triangle ABC with $a = 4.2$, $b = 2.6$ and angle $C = 43$ degrees.

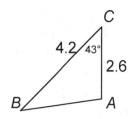

$$\text{Area} = \frac{1}{2}ab\sin C$$

$$= \frac{1}{2}(4.2)(2.6)\sin 43$$

$$= 3.724 \text{ square units}$$

1. Draw and label the sketch.

2. Write the formula.

3. Substitute.

4. Solve.

When only the lengths of the sides are known, it is possible to find the area of the triangle ABC using Heron's Formula:

$$\text{Area} = \sqrt{s(s-a)(s-b)(s-c)} \qquad \text{where } s = \frac{1}{2}(a+b+c)$$

Sample problem:

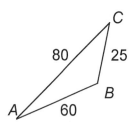

1. Draw and label the sketch.

First, find s.

$$s = \frac{1}{2}(a+b+c)$$

2. Write the Formula.

$$s = \frac{1}{2}(25 + 80 + 60)$$

3. Substitute.

$$s = 82.5$$

4. Solve.

Now find the area.

$$A = \sqrt{s(s-a)(s-b)(s-c)}$$

5. Write the formula.

$$A = \sqrt{82.5(82.5 - 25)(82.5 - 80)(82.5 - 60)}$$

6. Substitute.

$$A = 516.5616 \text{ square units}$$

7. Solve.

0012. Demonstrate an understanding of the fundamental concept of calculus.

The limit of a function is the y value that the graph approaches as the x values approach a certain number. To find a limit there are two points to remember.

1. Factor the expression completely and cancel all common factors in fractions.
2. Substitute the number to which the variable is approaching. In most cases this produces the value of the limit.

If the variable in the limit is approaching ∞, factor and simplify first; then examine the result. If the result does not involve a fraction with the variable in the denominator, the limit is usually also equal to ∞. If the variable is in the denominator of the fraction, the denominator is getting larger which makes the entire fraction smaller. In other words the limit is zero.

Examples:

1. $\displaystyle\lim_{x \to {}^-3} \frac{x^2 + 5x + 6}{x + 3} + 4x$ Factor the numerator.

   $\displaystyle\lim_{x \to {}^-3} \frac{(x+3)(x+2)}{(x+3)} + 4x$ Cancel the common factors.

   $\displaystyle\lim_{x \to {}^-3} (x+2) + 4x$ Substitute $^-3$ for x.

   $({}^-3 + 2) + 4({}^-3)$ Simplify.

   $^-1 + {}^-12$

   $^-13$

2. $\displaystyle\lim_{x \to \infty} \frac{2x^2}{x^5}$ Cancel the common factors.

   $\displaystyle\lim_{x \to \infty} \frac{2}{x^3}$ Since the denominator is getting larger, the entire fraction is getting smaller. The fraction is getting close to zero.

   $\dfrac{2}{\infty^3}$

Practice problems:

1. $\displaystyle\lim_{x \to \pi} 5x^2 + \sin x$ 2. $\displaystyle\lim_{x \to {}^-4} \frac{x^2 + 9x + 20}{x + 4}$

After simplifying an expression to evaluate a limit, substitute the value that the variable approaches. If the substitution results in either 0/0 or ∞/∞, use L'Hopital's rule to find the limit.

L'Hopital's rule states that you can find such limits by taking the derivative of the numerator and the derivative of the denominator, and then finding the limit of the resulting quotient.

Examples:

1. $\lim\limits_{x \to \infty} \dfrac{3x - 1}{x^2 + 2x + 3}$ — No factoring is possible.

$$\dfrac{3\infty - 1}{\infty^2 + 2\infty + 3}$$ — Substitute ∞ for x.

$$\dfrac{\infty}{\infty}$$ — Since a constant times infinity is still a large number, $3(\infty) = \infty$.

$$\lim\limits_{x \to \infty} \dfrac{3}{2x + 2}$$ — To find the limit, take the derivative of the numerator and denominator.

$$\dfrac{3}{2(\infty) + 2}$$ — Substitute ∞ for x again.

$$\dfrac{3}{\infty}$$ — Since the denominator is a very large number, the fraction is getting smaller. Thus the limit is zero.

$$0$$

2. $\lim\limits_{x \to 1} \dfrac{\ln x}{x - 1}$ Substitute 1 for x.

$\dfrac{\ln 1}{1 - 1}$ The $\ln 1 = 0$

$\dfrac{0}{0}$ To find the limit, take the derivative of the numerator and denominator.

$\lim\limits_{x \to 1} \dfrac{\frac{1}{x}}{1}$ Substitute 1 for x again.

$\dfrac{\frac{1}{1}}{1}$ Simplify. The limit is one.

1

Practice problems:

1. $\lim\limits_{x \to \infty} \dfrac{x^2 - 3}{x}$ 2. $\lim\limits_{x \to \frac{\pi}{2}} \dfrac{\cos x}{x - \dfrac{\pi}{2}}$

The **difference quotient** is the average rate of change over an interval. For a function f, the **difference quotient** is represented by the formula:

$$\frac{f(x + h) - f(x)}{h}$$

This formula computes the slope of the secant line through two points on the graph of f. These are the points with x-coordinates x and $x + h$.

<u>Example:</u> Find the difference quotient for the function $f(x) = 2x^2 + 3x - 5$.

$$\frac{f(x+h) - f(x)}{h} = \frac{2(x+h)^2 + 3(x+h) - 5 - (2x^2 + 3x - 5)}{h}$$

$$= \frac{2(x^2 + 2hx + h^2) + 3x + 3h - 5 - 2x^2 - 3x + 5}{h}$$

$$= \frac{2x^2 + 4hx + 2h^2 + 3x + 3h - 5 - 2x^2 - 3x + 5}{h}$$

$$= \frac{4hx + 2h^2 + 3h}{h}$$

$$= 4x + 2h + 3$$

The **derivative** is the slope of a tangent line to a graph $f(x)$, and is usually denoted $f'(x)$. This is also referred to as the instantaneous rate of change.

The derivative of $f(x)$ at $x = a$ is given by taking the limit of the average rates of change (computed by the difference quotient) as h approaches 0.

$$f'(a) = \lim_{h \to 0} \frac{f(a+h) - f(a)}{h}$$

Example: Suppose a company's annual profit (in millions of dollars) is represented by the above function $f(x) = 2x^2 + 3x - 5$ and x represents the number of years in the interval. Compute the rate at which the annual profit was changing over a period of 2 years.

$$f'(a) = \lim_{h \to 0} \frac{f(a+h) - f(a)}{h}$$

$$= f'(2) = \lim_{h \to 0} \frac{f(2+h) - f(2)}{h}$$

Using the difference quotient we computed above, $4x + 2h + 3$, we get

$$f'(2) = \lim_{h \to 0}(4(2) + 2h + 3)$$

$$= 8 + 3$$

$$= 11.$$

We have, therefore, determined that the annual profit for the company has increased at the average rate of $11 million per year over the two-year period.

The **definite integral** is defined as the limit approached by the *n*th upper and lower Riemann sums as $n \to \infty$. The **Riemann sum** is the sum of the areas of all rectangles approximating the area under the graph of a function.

Example: For $f(x) = x^2$, find the values of the Riemann Sums over the interval [0, 1] using *n* subintervals of equal width evaluated at the midpoint of each subinterval. Find the limit of the Riemann Sums.

$$\int_0^1 x^2\,dx$$

Let *n* be a positive integer and let $\displaystyle\lim_{n\to\infty} \frac{n(n+1)(2n+1)}{6n^3} = \frac{1}{3}$.

Take the interval [0,1] and subdivide it into *n* subintervals each of length $\dfrac{1}{n}$.

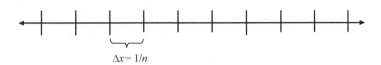

$\Delta x = 1/n$

Let $a_i = \dfrac{i}{n}$; the endpoints of the ith subinterval are

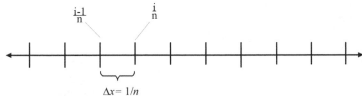

$\dfrac{i-1}{n}$ $\dfrac{i}{n}$

$\Delta x = 1/n$

Let $x_i = \dfrac{i}{n}$ be the right-hand endpoint. Draw a line of length

$f(x_i) = \left[\dfrac{i}{n}\right]^2$ at the right-hand endpoint.

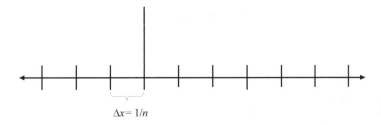

$\Delta x = 1/n$

Draw a rectangle.

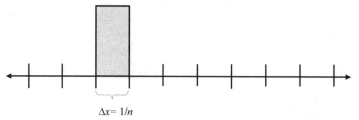

$\Delta x = 1/n$

The area of this rectangle is $f(x)\triangle x = \left[\dfrac{i}{n}\right]^2 \dfrac{1}{n} = \dfrac{i^2}{n^3}$.

Now draw all 9 rectangles.

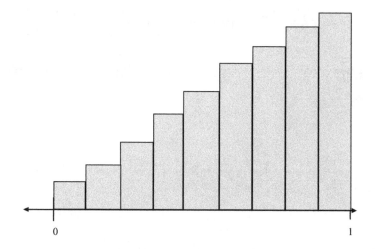

The sum of the area of these rectangles is:

$$\sum_{i=1}^{n} f(x_i)\triangle x = \sum_{i=1}^{n} \frac{i^2}{n^3} = \frac{1}{n^3}\sum_{i=1}^{n} i^2 = \frac{n(n+1)(2n+1)}{6n^3}.$$

Finally, to evaluate the integral, we take the limit

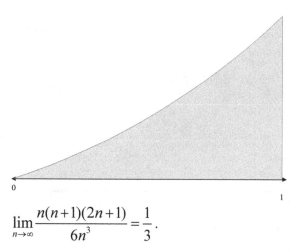

$$\lim_{n\to\infty} \frac{n(n+1)(2n+1)}{6n^3} = \frac{1}{3}.$$

An integral is almost the same thing as an antiderivative, the only difference is the notation.

$\int_{-2}^{1} 2x\,dx$ is the integral form of the antiderivative of $2x$. The numbers at the top and bottom of the integral sign (1 and $^-2$) are the numbers used to find the exact value of this integral. If these numbers are used the integral is said to be *definite* and does not have an unknown constant c in the answer.

The fundamental theorem of calculus states that an integral such as the one above is equal to the antiderivative of the function inside (here $2x$) evaluated from $x = {}^-2$ to $x = 1$. To do this, follow these steps.

1. Take the antiderivative of the function inside the integral.

2. Plug in the upper number (here $x = 1$) and plug in the lower number (here $x = {}^- 2$), giving two expressions.

3. Subtract the second expression from the first to achieve the integral value.

Examples:

1. $\int_{-2}^{1} 2x\,dx = x^2 \Big]_{-2}^{1}$ Take the antiderivative.

$\int_{-2}^{1} 2x\,dx = 1^2 - ({}^-2)^2$ Substitute in $x = 1$ and $x = {}^-2$ and subtract the results.

$\int_{-2}^{1} 2x\,dx = 1 - 4 = {}^- 3$ The integral has the value $^-3$.

2. $\int_{0}^{\pi/2} \cos x\,dx = \sin x \Big]_{0}^{\pi/2}$

     The antiderivative of $\cos x$ is $\sin x$.

$\int_{0}^{\frac{\pi}{2}} \cos x\,dx = \sin\frac{\pi}{2} - \sin 0$ Substitute in $x = \dfrac{\pi}{2}$ and $x = 0$.

     Subtract the results.

$\int_{0}^{\frac{\pi}{2}} \cos x\,dx = 1 - 0 = 1$ The integral has the value 1.

A list of integration formulas follows. In each case the letter u is used to represent either a single variable or an expression. Note that also in each case du is required. du is the derivative of whatever u stands for. If u is sin x then du is cos$x$, which is the derivative of sin$x$.

If the derivative of u is not entirely present, remember you can put in constants as long as you also insert the reciprocal of any such constants. n is a natural number.

$$\int u^n du = \frac{1}{n+1}u^{n+1} + c \quad \text{if } n \neq {}^-1$$

$$\int \frac{1}{u} du = \ln|u| + c$$

$$\int e^u du = e^u + c$$

$$\int a^u du = \frac{1}{\ln a}a^u + c$$

$$\int \sin u\, du = {}^-\cos u + c$$

$$\int \cos u\, du = \sin u + c$$

$$\int \sec^2 u\, du = \tan u + c$$

$$\int \csc^2 u\, du = {}^-\cot u + c$$

Example:

1. $\int \frac{6}{x} dx = 6\int \frac{1}{x} dx$

   You can pull any constants outside the integral.

   $\int \frac{6}{x} dx = 6\ln|x| + c$

Sometimes an integral does not always look exactly like the forms from the previous page. But with a simple substitution (sometimes called a *u* substitution), the integral can be made to look like one of the general forms.

You might need to experiment with different *u* substitutions before you find the one that works. Follow these steps.

1. If the object of the integral is a sum or difference, first split the integral up.
2. For each integral, see if it fits one of the general forms from the previous pg.
3. If the integral does not fit one of the forms, substitute the letter *u* in place of one of the expressions in the integral.
4. Off to the side, take the derivative of *u*, and see if that derivative exists inside the original integral. If it does, replace that derivative expression with *du*. If it does not, try another *u* substitution.
5. Now the integral should match one of the general forms, including both the *u* and the *du*.
6. Take the integral using the general forms, and substitute for the value of *u*.

Examples:

1. $\int \left(\sin x^2 \bullet 2x + \cos x^2 \bullet 2x \right) dx$ Split the integral up.

$\int \sin(x^2) \bullet 2x\,dx + \int \cos(x^2)2x\,dx$

$u = x^2, \quad du = 2x\,dx$ If you let $u = x^2$, the derivative of *u, du, is* $2x\,dx$.

$\int \sin u\,du + \int \cos u\,du$ Make the *u* and *du* substitutions.

$^-\cos u + \sin u + c$ Use the formula for integrating sin*u* and cos*u*.

$^-\cos(x^2) + \sin(x^2) + c$ Substitute back in for *u*.

2. $\int e^{\sin x} \cos x\, dx$

$u = \cos x,\ du = {}^{-}\sin x\, dx$

Try letting $u = \cos x$. The derivative of u would be ${}^{-}\sin x\, dx$, which is not present.

$u = \sin x,\ du = \cos x\, dx$

Try another substitution: $u = \sin x$, $du = \cos x\, dx$. $du = \cos x\, dx$ is present.

$\int e^u\, du$

e^u is one of the general forms.

e^u

The integral of e^u is e^u.

$e^{\sin x}$

Substitute back in for u.

Integration by parts should only be used if none of the other integration methods works. Integration by parts requires two substitutions (both u and dv).

1. Let dv be the part of the integral that you think can be integrated by itself.
2. Let u be the part of the integral which is left after the dv substitution is made.
3. Integrate the dv expression to arrive at just simply v.
4. Differentiate the u expression to arrive at du. If u is just x, then du is dx.
5. Rewrite the integral using $\int u\, dv = uv - \int v\, du$.
6. All that is left is to integrate $\int v\, du$.
7. If you cannot integrate $v\, du$, try a different set of substitutions and start the process over.

Examples:

1. $\int x e^{3x}\, dx$

Make dv and u substitutions.

$dv = e^{3x} dx \qquad u = x$

Integrate the dv term to arrive at v.

$v = \dfrac{1}{3} e^{3x} \qquad du = dx$

Differentiate the u term to arrive at du.

$$\int xe^{3x}dx = x\left(\frac{1}{3}e^{3x}\right) - \int \frac{1}{3}e^{3x}dx$$

Rewrite the integral using the above formula. Before taking the integral of $\frac{1}{3}e^{3x}dx$, you must put in a 3 and another $1/3$.

$$\int xe^{3x}dx = \frac{1}{3}xe^{3x} - \frac{1}{3}\bullet\frac{1}{3}\int e^{3x}3dx$$

$$\int xe^{3x}dx = \frac{1}{3}xe^{3x} - \frac{1}{9}e^{3x} + c$$

Integrate to arrive at the solution.

2. $\int \ln 4x\,dx$

Note that no other integration method will work.

$dv = dx \qquad u = \ln 4x$

Make the dv and u substitutions.

$v = x \qquad du = \frac{1}{4x}\bullet 4 = \frac{1}{x}dx$

Integrate dx to get x.

Differentiate $\ln 4x$ to get $(1/x)\,dx$.

$$\int \ln 4x\,dx = \ln 4x \bullet x - \int x \bullet \frac{1}{x}dx$$

Rewrite the formula above.

$$\int \ln 4x\,dx = \ln 4x \bullet x - \int dx$$

Simplify the integral.

$$\int \ln 4\,dx = \ln 4x \bullet x - x + c$$

Integrate dx to get the value $x + c$

The **sums of terms in a progression** is simply found by determining if it is an arithmetic or geometric sequence and then using the appropriate formula.

Sum of first n terms of an arithmetic sequence.

$$S_n = \frac{n}{2}(a_1 + a_n)$$

or

$$S_n = \frac{n}{2}\left[2a_1 + (n-1)d\right]$$

$$S_n = \frac{a_1\left(r^n - 1\right)}{r - 1}, r \neq 1$$

Sum of first n terms of a geometric sequence.

Sample Problems:

1. $\displaystyle\sum_{i=1}^{10}(2i+2)$

    This means find the sum of the terms beginning with the first term and ending with the 10th term of the sequence $a = 2i + 2$.

    $a_1 = 2(1) + 2 = 4$

    $a_{10} = 2(10) + 2 = 22$

    $S_n = \dfrac{n}{2}(a_1 + a_n)$

    $S_n = \dfrac{10}{2}(4 + 22)$

    $S_n = 130$

2. Find the sum of the first 6 terms in an arithmetic sequence if the first term is 2 and the common difference d, is -3.

    $n = 6 \qquad a_1 = 2 \qquad d = {}^-3$

    $S_n = \dfrac{n}{2}\left[2a_1 + (n-1)d\right]$

    $S_6 = \dfrac{6}{2}\left[2 \times 2 + (6-1)\,{}^-3\right]$ \qquad Substitute known values.

    $S_6 = 3\left[4 + \left({}^-15\right)\right]$ \qquad Solve.

    $S_6 = 3(-11) = -33$

3. Find $\displaystyle\sum_{i=1}^{5}4 \times 2^{\,i}$

    This means the sum of the first 5 terms where $a_i = a \times b^i$ and $r = b$.

    $a_1 = 4 \times 2^1 = 8$ \qquad Identify a_1, r, n

    $r = 2 \qquad n = 5$

    $S_n = \dfrac{a_1(r^n - 1)}{r - 1}$ \qquad Substitute a, r, n

    $S_5 = \dfrac{8(2^5 - 1)}{2 - 1}$ \qquad Solve.

    $S_5 = \dfrac{8(31)}{1} = 248$

Practice problems:

1. Find the sum of the first five terms of the sequence if $a = 7$ and $d = 4$.

2. $\displaystyle\sum_{i=1}^{7}(2i - 4)$

3. $\displaystyle\sum_{i=1}^{6} {}^{-}3\left(\frac{2}{5}\right)^{i}$

To find the **slope of a curve at a point**, there are two steps to follow.

      1.   Take the derivative of the function.
      2.   Plug in the value to find the slope.

If plugging into the derivative yields a value of zero, the tangent line is horizontal at that point.

If plugging into the derivative produces a fraction with zero in the denominator, the tangent line at this point has an undefined slope and is thus a vertical line.

Examples:

1. Find the slope of the tangent line for the given function at the given point.

$$y = \frac{1}{x-2} \quad \text{at } (3,1)$$

$y = (x-2)^{-1}$ Rewrite using negative exponents.

$\dfrac{dy}{dx} = {}^{-}1(x-2)^{-1-1}(1)$ Use the Chain rule.

                             The derivative of $(x-2)$ is 1.

$\dfrac{dy}{dx} = {}^{-}1(x-2)^{-2}$

$\dfrac{dy}{dx}\Big|_{x=3} = {}^{-}1(3-2)^{-2}$ Evaluate at $x = 3$.

$\dfrac{dy}{dx}\Big|_{x=3} = {}^{-}1$ The slope of the tangent line is ${}^{-}1$ at $x = 3$.

2. Find the points where the tangent to the curve $f(x) = 2x^2 + 3x$ is parallel to the line $y = 11x - 5$.

$f'(x) = 2 \bullet 2x^{2-1} + 3$	Take the derivative of $f(x)$ to get the slope of a tangent line.
$f'(x) = 4x + 3$	
$4x + 3 = 11$	Set the slope expression $(4x + 3)$ equal to the slope of $y = 11x - 5$.
$x = 2$	Solve for the x value of the point.
$f(2) = 2(2)^2 + 3(2)$	The y value is 14.
$f(2) = 14$	So $(2, 14)$ is the point on $f(x)$ where the tangent line is parallel to $y = 11x - 5$.

To **write an equation** of a tangent line at a point, two things are needed.

A point--the problem will usually provide a point, (x, y). If the problem only gives an x value, plug the value into the original function to get the y coordinate.

The slope--to find the slope, take the derivative of the original function. Then plug in the x value of the point to get the slope.

After obtaining a point and a slope, use the Point-Slope form for the equation of a line:

$$(y - y_1) = m(x - x_1)$$

where m is the slope and (x_1, y_1) is the point.

Example:

Find the equation of the tangent line to $f(x) = 2e^{x^2}$ at $x = {}^-1$.

$f({}^-1) = 2e^{({}^-1)^2}$ Plug in the x coordinate to obtain the y coordinate.

$= 2e^1$ The point is $({}^-1, 2e)$.

$f'(x) = 2e^{x^2} \bullet (2x)$

$f'({}^-1) = 2e^{({}^-1)^2} \bullet (2 \bullet {}^-1)$

$f'({}^-1) = 2e^1({}^-2)$

$f'({}^-1) = {}^-4e$ The slope at $x = {}^-1$ is ${}^-4e$.

$(y - 2e) = {}^-4e(x - {}^-1)$ Plug in the point $({}^-1, 2e)$ and the slope $m = {}^-4e$. Use the point slope form of a line.

$y = {}^-4ex - 4e + 2e$

$y = {}^-4ex - 2e$ Simplify to obtain the equation for the tangent line.

A **normal line** is a line which is perpendicular to a tangent line at a given point. Perpendicular lines have slopes which are negative reciprocals of each other. To find the equation of a normal line, first get the slope of the tangent line at the point. Find the negative reciprocal of this slope. Next, use the new slope and the point on the curve, both the x_1 and y_1 coordinates, and substitute into the Point-Slope form of the equation for a line:

$$(y - y_1) = slope \bullet (x - x_1)$$

Examples:

1. Find the equation of the normal line to the tangent to the curve $y = (x^2 - 1)(x - 3)$ at $x = {}^-2$.

$f(-2) = (({}^-2)^2 - 1)({}^-2 - 3)$ First find the y coordinate of the point on the curve. Here,

$f(-2) = -15$ $y = {}^-15$ when $x = {}^-2$.

$y = x^3 - 3x^2 - x + 3$ Before taking the derivative, multiply the expression first. The derivative of a sum is easier to find than the derivative of a product.

$$y' = 3x^2 - 6x - 1$$

Take the derivative to find the slope of the tangent line.

$$y'_{x=^-2} = 3(^-2)^2 - 6(^-2) - 1$$
$$y'_{x=^-2} = 23$$

$$\text{slope of normal} = \frac{^-1}{23}$$

For the slope of the normal line, take the negative reciprocal of the tangent line's slope.

$$(y - {}^-15) = \frac{^-1}{23}(x - {}^- 2)$$

Plug (x_1, y_1) into the point-slope equation.

$$(y + 15) = \frac{^-1}{23}(x + 2)$$

$$y = -\frac{1}{23}x - 14\frac{21}{23}$$

$$y = -\frac{1}{23}x + \frac{2}{23} - 15 = \frac{1}{23}x - 14\frac{21}{23}$$

2. Find the equation of the normal line to the tangent to the curve $y = \ln(\sin x)$ at $x = \pi$.

$$f(\pi) = \ln(\sin \pi)$$

$\sin \pi = 1$ and $\ln(1) = 0$ (recall $e^0 = 1$).
So $x_1 = \pi$ and $y_1 = 0$.

$$f(\pi) = \ln(1) = 0$$

$$y' = \frac{1}{\sin x} \bullet \cos x$$

Take the derivative to find the slope of the tangent line.

$$y'_{x=\pi} = \frac{\cos \pi}{\sin \pi} = \frac{0}{1}$$

Slope of normal does not exist.

$\dfrac{^-1}{0}$ does not exist. So the normal line is vertical at $x = \pi$.

0013. Apply the principles and techniques of calculus to model and solve problems.

The derivative of a function has two basic interpretations.

I. Instantaneous rate of change
II. Slope of a tangent line at a given point

If a question asks for the rate of change of a function, take the derivative to find the equation for the rate of change. Then plug in for the variable to find the instantaneous rate of change.

The following is a list summarizing some of the more common quantities referred to in rate of change problems.

area	height	profit
decay	population growth	sales
distance	position	temperature
frequency	pressure	volume

Pick a point, say $x = {}^-3$, on the graph of a function. Draw a tangent line at that point. Find the derivative of the function and plug in $x = {}^-3$. The result will be the slope of the tangent line.

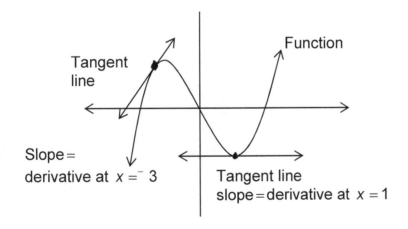

Tangent line

Function

Slope = derivative at $x = {}^-3$

Tangent line slope = derivative at $x = 1$

Extreme value problems are also known as max-min problems. Extreme value problems require using the first derivative to find values which either maximize or minimize some quantity such as area, profit, or volume. Follow these steps to solve an extreme value problem.

1. Write an equation for the quantity to be maximized or minimized.
2. Use the other information in the problem to write secondary equations.
3. Use the secondary equations for substitutions, and rewrite the original equation in terms of only one variable.
4. Find the derivative of the primary equation (step 1) and the critical values of this derivative.
5. Substitute these critical values into the primary equation.

The value which produces either the largest or smallest value is used to find the solution.

Example:
A manufacturer wishes to construct an open box from the piece of metal shown below by cutting squares from each corner and folding up the sides. The square piece of metal is 12 feet on a side. What are the dimensions of the squares to be cut out which will maximize the volume?

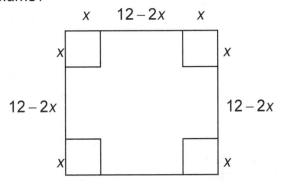

Volume $= lwh$		Primary equation.

$l = 12 - 2x$

$w = 12 - 2x$ Secondary equations.

$h = x$

$V = (12 - 2x)(12 - 2x)(x)$ Make substitutions.

$V = (144x - 48x^2 + 4x^3)$ Take the derivative.

$\dfrac{dV}{dx} = 144 - 96x + 12x^2$

      Find critical values by setting the derivative equal to zero.

$0 = 12(x^2 - 8x + 12)$

$0 = 12(x - 6)(x - 2)$

$x = 6$ and $x = 2$ Substitute critical values into volume equation.

$V = 144(6) - 48(6)^2 + 4(6)^3$ $V = 144(2) - 48(2)^2 + 4(2)^3$

$V = 0$ ft^3 when $x = 6$ $V = 128$ ft^3 when $x = 2$

Therefore, the manufacturer can maximize the volume if the squares to be cut out are 2 feet by 2 feet ($x = 2$).

If a particle (or a car, a bullet, etc.) is moving along a line, then the distance that the particle travels can be expressed by a function in terms of time.

1. The first derivative of the distance function will provide the velocity function for the particle. Substituting a value for time into this expression will provide the instantaneous velocity of the particle at the time. Velocity is the rate of change of the distance traveled by the particle. Taking the absolute value of the derivative provides the speed of the particle. A positive value for the velocity indicates that the particle is moving forward, and a negative value indicates the particle is moving backwards.

2. The second derivative of the distance function (which would also be the first derivative of the velocity function) provides the acceleration function. The acceleration of the particle is the rate of change of the velocity. If a value for time produces a positive acceleration, the particle is speeding up; if it produces a negative value, the particle is slowing down. If the acceleration is zero, the particle is moving at a constant speed.

To find the time when a particle stops, set the first derivative (i.e. the velocity function) equal to zero and solve for time. This time value is also the instant when the particle changes direction.

Example:

The motion of a particle moving along a line is according to the equation:

$s(t) = 20 + 3t - 5t^2$ where s is in meters and t is in seconds. Find the position, velocity, and acceleration of a particle at $t = 2$ seconds.

$s(2) = 20 + 3(2) - 5(2)^2$ $\quad = 6$ meters	Plug $t = 2$ into the original equation to find the position.
$s\,'(t) = v(t) = 3 - 10t$	The derivative of the first function gives the velocity.
$v(2) = 3 - 10(2) = \,^-17\,\text{m/s}$	Plug $t = 2$ into the velocity function to find the velocity. $^-17\,\text{m/s}$ indicates the particle is moving backwards.
$s\,''(t) = a(t) = \,^-10$	The second derivation of position gives the acceleration. Substitute
$a(2) = \,^-10\,\text{m/s}$	$t = 2$, yields an acceleration of $^-10\,\text{m/s}$, which indicates the particle is slowing down.

Finding the **rate of change** of one quantity (for example distance, volume, etc.) with respect to time it is often referred to as a rate of change problem. To find an instantaneous rate of change of a particular quantity, write a function in terms of time for that quantity; then take the derivative of the function. Substitute in the values at which the instantaneous rate of change is sought.

Functions which are in terms of more than one variable may be used to find related rates of change. These functions are often not written in terms of time. To find a related rate of change, follow these steps.

1. Write an equation which relates all the quantities referred to in the problem.

2. Take the derivative of both sides of the equation with respect to time.

   Follow the same steps as used in implicit differentiation. This means take the derivative of each part of the equation remembering to multiply each term by the derivative of the variable involved with respect to time. For example, if a term includes the variable v for volume, take the derivative of the term remembering to multiply by dv/dt for the derivative of volume with respect to time. dv/dt is the rate of change of the volume.

3. Substitute the known rates of change and quantities, and solve for the desired rate of change.

Example:

1. What is the instantaneous rate of change of the area of a circle where the radius is 3 cm?

$A(r) = \pi r^2$	Write an equation for area.
$A'(r) = 2\pi r$	Take the derivative to find the rate of change.
$A'(3) = 2\pi(3) = 6\pi$	Substitute in $r = 3$ to arrive at the instantaneous rate of change.

The **derivative of a distance function** provides a velocity function, and the derivative of a velocity function provides an acceleration function. Therefore taking the antiderivative of an acceleration function yields the velocity function, and the antiderivative of the velocity function yields the distance function.

Example:

A particle moves along the x axis with acceleration $a(t) = 3t - 1$ cm/sec/sec. At time $t = 4$, the particle is moving to the left at 3 cm per second. Find the velocity of the particle at time $t = 2$ seconds.

$a(t) = 3t - 1$ Before taking the antiderivative, make sure the correct coefficients are present.

$$a(t) = 3 \bullet \frac{1}{2} \bullet 2t - 1$$

$v(t) = \frac{3}{2}t^2 - 1 \bullet t + c$ $v(t)$ is the antiderivative of $a(t)$.

$$v(4) = \frac{3}{2}(4)^2 - 1(4) + c = {}^-3$$

Use the given information that $v(4) = {}^-3$ to find c.

$24 - 4 + c = {}^- 3$

$20 + c = {}^-3$

$c = {}^-23$ The constant is $^-23$.

$v(t) = \frac{3}{2}t^2 - 1t + {}^-23$ Rewrite $v(t)$ using $c = {}^-23$.

$v(2) = \frac{3}{2}2^2 - 1(2) + {}^-23$ Solve $v(t)$ at $t=2$.

$v(2) = 6 + 2 + {}^-23 = {}^-15$ the velocity at $t = 2$ is -15 cm/sec.

Practice problem:

A particle moves along a line with acceleration $a(t) = 5t + 2$. The velocity after 2 seconds is $^-10$ m/sec.

1. Find the initial velocity.
2. Find the velocity at $t = 4$.

To **find the distance function**, take the antiderivative of the velocity function. And to find the velocity function, find the antiderivative of the acceleration function. Use the information in the problem to solve for the constants that result from taking the antiderivatives.

Example:
A particle moves along the x axis with acceleration $a(t) = 6t - 6$. The initial velocity is 0 m/sec and the initial position is 8 cm to the right of the origin. Find the velocity and position functions.

$v(0) = 0$	Interpret the given information.
$s(0) = 8$	
$a(t) = 6t - 6$	Put in the coefficients needed to take the antiderivative.
$a(t) = 6 \bullet \dfrac{1}{2} \bullet 2t - 6$	
$v(t) = \dfrac{6}{2}t^2 - 6t + c$	Take the antiderivative of $a^{(t)}$ to get $v(t)$.
$v(0) = 3(0)^2 - 6(0) + c = 0$	Use $v(0) = 0$ to solve for c.
$0 - 0 + c = 0$	
$\quad c = 0$	$c = 0$
$v(t) = 3t^2 - 6t + 0$	Rewrite $v(t)$ using $c = 0$.
$v(t) = 3t^2 - 6\dfrac{1}{2} \bullet 2t$	Put in the coefficients needed to take the antiderivative.
$s(t) = t^3 - \dfrac{6}{2}t^2 + c$	Take the antiderivative of $v^{(t)}$ to get $s(t) \rightarrow$ the distance function.
$s(0) = 0^3 - 3(0)^2 + c = 8$	Use $s(0) = 8$ to solve for c.
	$c = 8$
$s(t) = t^3 - 3t^2 + 8$	

A. Derivative of a constant--for any constant, the derivative is always zero.

B. Derivative of a variable--the derivative of a variable (i.e. x) is one.

C. Derivative of a variable raised to a power--for variable expressions with rational exponents (i.e. $3x^2$) multiply the coefficient (3) by the exponent (2) then subtract one (1) from the exponent to arrive at the derivative $\left(3x^2\right) = \left(6x\right)$

Example:

1. $y = 5x^4$ Take the derivative.

$$\frac{dy}{dx} = (5)(4)x^{4-1}$$

Multiply the coefficient by the exponent and subtract 1 from the exponent.

$$\frac{dy}{dx} = 20x^3$$ Simplify.

2. $y = \dfrac{1}{4x^3}$ Rewrite using negative exponent.

$$y = \frac{1}{4}x^{-3}$$ Take the derivative.

$$\frac{dy}{dx} = \left(\frac{1}{4} \bullet {}^-3\right)x^{-3-1}$$

$$\frac{dy}{dx} = \frac{{}^-3}{4}x^{-4} = \frac{{}^-3}{4x^4}$$ Simplify.

3. $y = 3\sqrt{x^5}$ Rewrite using $\sqrt[z]{x^n} = x^{n/z}$.

$y = 3x^{5/2}$ Take the derivative.

$$\frac{dy}{dx} = (3)\left(\frac{5}{2}\right)x^{5/2-1}$$

$$\frac{dy}{dx} = \left(\frac{15}{2}\right)x^{3/2}$$ Simplify.

$$\frac{dy}{dx} = 7.5\sqrt{x^3} = 7.5x\sqrt{x}$$

A. sin x -- the derivative of the sine of x is simply the cosine of x.

B. cos x --the derivative of the cosine of x is negative one ($^{-}1$) times the sine of x.

C. tan x -- the derivative of the tangent of x is the square of the secant of x.

If object of the trig. function is an expression other than x, follow the above rules substituting the expression for x. The only additional step is to multiply the result by the derivative of the expression.

Examples:

1. $y = \pi \sin x$

$\dfrac{dy}{dx} = \pi \cos x$

Carry the coefficient (π) throughout the problem.

2. $y = \dfrac{2}{3}\cos x$

Do not forget to multiply the coefficient by negative one when taking the derivative of a cosine function.

$\dfrac{dy}{dx} = \dfrac{^{-}2}{3}\sin x$

3. $y = 4\tan\left(5x^3\right)$

$\dfrac{dy}{dx} = 4\sec^2\left(5x^3\right)\left(5 \bullet 3x^{3-1}\right)$

The derivative of $\tan x$ is $\sec^2 x$.

$\dfrac{dy}{dx} = 4\sec^2\left(5x^3\right)\left(15x^2\right)$

Carry the $\left(5x^3\right)$ term throughout the problem.

$\dfrac{dy}{dx} = 4 \bullet 15x^2 \sec^2\left(5x^3\right)$

Multiply $4\sec^2\left(5x^3\right)$ by the derivative of $5x^3$.

$\dfrac{dy}{dx} = 60x^2 \sec^2\left(5x^3\right)$

Rearrange the terms and simplify.

$f(x) = e^x$ is an exponential function. The derivative of e^x is exactly the same thing$\rightarrow e^x$. If instead of x, the exponent on e is an expression, the derivative is the same e raised to the algebraic exponent multiplied by the derivative of the algebraic expression.

If a base other than e is used, the derivative is the natural log of the base times the original exponential function times the derivative of the exponent.

Examples:

1. $y = e^x$

$$\frac{dy}{dx} = e^x$$

2. $y = e^{3x}$

$$\frac{dy}{dx} = e^{3x} \bullet 3 = 3e^{3x}$$

Multiply e^{3x} by the derivative of $3x$ which is 3.

$$\frac{dy}{dx} = 3e^{3x}$$

Rearrange the terms.

3. $y = \dfrac{5}{e^{\sin x}}$

$y = 5e^{-\sin x}$

Rewrite using negative exponents.

$$\frac{dy}{dx} = 5e^{-\sin x} \bullet \left(^{-}\cos x\right)$$

Multiply $5e^{-\sin x}$ by the derivative of $^{-}\sin x$ which is $^{-}\cos x$.

$$\frac{dy}{dx} = \frac{^{-}5\cos x}{e^{\sin x}}$$

Use the definition of negative exponents to simplify.

4. $y = {}^{-}2 \bullet \ln 3^{4x}$

$$\frac{dy}{dx} = {}^{-}2 \bullet (\ln 3)\left(3^{4x}\right)(4)$$

The natural log of the base is ln3.

The derivative of $4x$ is 4.

$$\frac{dy}{dx} = {}^{-}8 \bullet 3^{4x} \ln 3$$

Rearrange terms to simplify

The most common logarithmic function on the Exam is the natural logarithmic function ($\ln x$). The derivative of $\ln x$ is simply $1/x$. If x is replaced by an algebraic expression, the derivative is the fraction one divided by the expression multiplied by the derivative of the expression.

For all other logarithmic functions, the derivative is 1 over the argument of the logarithm multiplied by 1 over the natural logarithm (ln) of the base multiplied by the derivative of the argument. Examples:

1. $y = \ln x$

$$\frac{dy}{dx} = \frac{1}{x}$$

2. $y = 3\ln\left(x^{-2}\right)$

$$\frac{dy}{dx} = 3 \bullet \frac{1}{x^{-2}} \bullet \left(^-2x^{-2-1}\right)$$

Multiply one over the argument (x^{-2}) by the derivative of x^{-2} which is $^-2x^{-2-1}$.

$$\frac{dy}{dx} = 3 \bullet x^2 \bullet \left(^-2x^{-3}\right)$$

$$\frac{dy}{dx} = \frac{^-6x^2}{x^3}$$

Simplify using the definition of negative exponents.

$$\frac{dy}{dx} = \frac{^-6}{x}$$

Cancel common factors to simplify.

3. $y = \log_5(\tan x)$

$$\frac{dy}{dx} = \frac{1}{\tan x} \bullet \frac{1}{\ln 5} \bullet (\sec^2 x)$$

The derivative of $\tan x$ is $\sec^2 x$.

$$\frac{dy}{dx} = \frac{\sec^2 x}{(\tan x)(\ln 5)}$$

Implicitly defined functions are ones where both variables (usually x and y) appear in the function. All of the rules for finding the derivative still apply to both variables. The only difference is that, while the derivative of x is one (1) and typically not even mentioned, the derivative of y must be written y' or dy/dx. Work these problems just like the other derivative problems, just remember to include the extra step of multiplying by dy/dx in the result.

If the question asks for the derivative of y, given an equation with x and y on both sides, find the derivative of each side. Then solve the new equation for dy/dx just as you would an algebra problem.

Examples:

1. $\dfrac{d}{dx}\left(y^3\right) = 3y^{3-1} \cdot \dfrac{dy}{dx}$

    Recall the derivative of x^3 is $3x^{3-1}$. Follow the same rule, but also multiply by the derivative of y which

    $\dfrac{d}{dx}(y^3) = 3y^2 \dfrac{dy}{dx}$ is dy/dx.

2. $\dfrac{d}{dx}(3\ln y) = 3 \cdot \dfrac{1}{y} \cdot \dfrac{dy}{dx}$

3. $\dfrac{d}{dx}(^-2\cos y) = ^-2(^-1\sin y)\dfrac{dy}{dx}$

    Recall the derivative of $\cos x$ is $^-\sin x$.

    $\dfrac{d}{dx}(^-2\cos y) = 2\sin y \dfrac{dy}{dx}$

4. $2y = e^{3x}$

    Solve for after taking the derivative.

    $2 \cdot \dfrac{dy}{dx} = e^{3x} \cdot 3$

    The derivative of e^{3x} is $e^{3x} \cdot 3$.

    $\dfrac{dy}{dx} = \dfrac{3}{2}e^{3x}$

    Divide both sides by 2 to solve for the derivative dy/dx.

A.   Derivative of a sum--find the derivative of each term separately and add the results.

B.   Derivative of a product--multiply the derivative of the first factor by the second factor and add to it the product of the first factor and the derivative of the second factor.

Remember the phrase "first times the derivative of the second plus the second times the derivative of the first."

C.   Derivative of a quotient--use the rule "bottom times the derivative of the top minus the top times the derivative of the bottom all divided by the bottom squared."

Examples:

1. $y = 3x^2 + 2\ln x + 5\sqrt{x}$ $\sqrt{x} = x^{1/2}$.

$\dfrac{dy}{dx} = 6x^{2-1} + 2 \cdot \dfrac{1}{x} + 5 \cdot \dfrac{1}{2}x^{1/2-1}$

$\dfrac{dy}{dx} = 6x + \dfrac{2}{x} + \dfrac{5}{2} \cdot \dfrac{1}{\sqrt{x}} = \dfrac{12x^2 + 4 + 5\sqrt{x}}{2x}$

$x^{1/2-1} = x^{-1/2} = 1/\sqrt{x}$.

$= \dfrac{12x^2 + 5\sqrt{x} + 4}{2x}$

2. $y = 4e^{x^2} \bullet \sin x$

$\dfrac{dy}{dx} = 4(e^{x^2} \bullet \cos x + \sin x \bullet e^{x^2} \bullet 2x)$

The derivative of e^{x^2} is $e^{x^2} \bullet 2$.

$\dfrac{dy}{dx} = 4(e^{x^2} \cos x + 2xe^{x^2} \sin x)$

$\dfrac{dy}{dx} = 4e^{x^2} \cos x + 8xe^{x^2} \sin x$

3. $y = \dfrac{\cos x}{x}$

$\dfrac{dy}{dx} = \dfrac{x(^-\sin x) - \cos x \bullet 1}{x^2}$

The derivative of x is 1.
The derivative of $\cos x$ is $^-\sin x$.

$\dfrac{dy}{dx} = \dfrac{^-x \sin x - \cos x}{x^2}$

A **composite function** is made up of two or more separate functions such as $\sin(\ln x)$ or $x^2 e^{3x}$. To find the derivatives of these composite functions requires two steps. First identify the predominant function in the problem. For example, in $\sin(\ln x))$ the major function is the sine function. In $x^2 e^{3x}$ the major function is a product of two expressions (x^2 and e^{3x}). Once the predominant function is identified, apply the appropriate differentiation rule. Be certain to include the step of taking the derivative of every part of the functions which comprise the composite function. Use parentheses as much as possible.

Examples:

1. $y = \sin(\ln x)$ The major function is a sine function.

$$\frac{dy}{dx} = \left[\cos(\ln x)\right] \bullet \left[\frac{1}{x}\right]$$ The derivative of $\sin x$ is $\cos x$.

The derivative of $\ln x$ is $1/x$.

2. $y = x^2 \bullet e^{3x}$ The major function is a product.

$$\frac{dy}{dx} = x^2 \left(e^{3x} \bullet 3\right) + e^{3x} \bullet 2x$$

The derivative of a product is "First $\frac{dy}{dx} = 3x^2 e^{3x} + 2xe^{3x}$ times the derivative of second plus the second times the derivative of the first."

3. $y = \tan^2\left(\frac{\ln x}{\cos x}\right)$

This function is made of several functions. The major function is a power function.

$$\frac{dy}{dx} = \left[2\tan^{2-1}\left(\frac{\ln x}{\cos x}\right)\right]\left[\sec^2\left(\frac{\ln x}{\cos x}\right)\right]\left[\frac{d}{dx}\left(\frac{\ln x}{\cos x}\right)\right]$$

The derivative of $\tan x$ is $\sec^2 x$. Hold off one more to take the derivative of $\ln x/\cos x$.

$$\frac{dy}{dx} = \left[2\tan\left(\frac{\ln x}{\cos x}\right)\sec^2\left(\frac{\ln x}{\cos x}\right)\right]\left[\frac{(\cos x)(1/x) - \ln x(^-\sin x)}{\cos^2 x}\right]$$

$$\frac{dy}{dx} = \left[2\tan\left(\frac{\ln x}{\cos x}\right)\sec^2\left(\frac{\ln x}{\cos x}\right)\right]\left[\frac{(\cos x)(1/x) + \ln x(\sin x)}{\cos^2 x}\right]$$

The derivative of a quotient is "Bottom times the derivative of the top minus the top times the derivative of the bottom all divided by the bottom squared."

If a question simply asks for the **derivative of a function**, the question is asking for the first derivative. To find the second derivative of a function, take the derivative of the first derivative. To find the third derivative, take the derivative of the second derivative; and so on. All of the regular derivative rules still apply.

Examples:

1. Find the second derivative $\left(\dfrac{d^2y}{dx^2} \text{ or } y''\right)$ of the following function:

$$y = 5x^2$$

$\dfrac{dy}{dx} = 2 \bullet 5x^{2-1} = 10$ Take the first derivative.

$\dfrac{d^2y}{dx^2} = 10$ The derivative of $10x$ is 10.

2. Find the third derivative (y''') of $f(x) = 4x^{3/2}$:

$$y' = \left(4 \bullet \frac{3}{2}\right)x^{\left(\frac{3}{2}-1\right)} = 6x^{\frac{1}{2}}$$

$$y'' = \left(6 \bullet \frac{1}{2}\right)x^{\left(\frac{1}{2}-1\right)} = 3x^{-\frac{1}{2}}$$

$$y''' = \left[3 - \left(\frac{1}{2}\right)\right]x^{\left(-\frac{1}{2}-1\right)} = -\frac{3}{2}x^{-\frac{3}{2}}$$

A function is said to be increasing if it is rising from left to right and decreasing if it is falling from left to right. Lines with positive slopes are increasing, and lines with negative slopes are decreasing. If the function in question is something other than a line, simply refer to the slopes of the tangent lines as the test for increasing or decreasing. Take the derivative of the function and plug in an x value to get the slope of the tangent line; a positive slope means the function is increasing and a negative slope means it is decreasing. If an interval for x values is given, just pick any point between the two values to substitute.

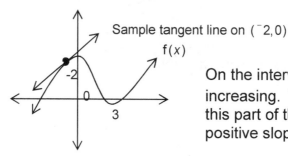

Sample tangent line on $(^-2,0)$

$f(x)$

On the interval $(^-2,0)$, $f(x)$ is increasing. The tangent lines on this part of the graph have positive slopes.

Example:

The growth of a certain bacteria is given by $f(x) = x + \dfrac{1}{x}$. Determine if the rate of growth is increasing or decreasing on the time interval $(^-1,0)$.

$$f'(x) = 1 + \frac{^-1}{x^2}$$

To test for increasing or decreasing, find the slope of the tangent line by taking the derivative.

$$f'\left(\frac{^-1}{2}\right) = 1 + \frac{^-1}{(^-1/2)^2}$$

Pick any point on $(^-1,0)$ and substitute into the derivative.

$$f'\left(\frac{^-1}{2}\right) = 1 + \frac{^-1}{1/4}$$

$$= 1 - 4$$

$$= ^-3$$

The slope of the tangent line at $x = \dfrac{^-1}{2}$ is $^-3$. The exact value of the slope is not important. The important fact is that the slope is negative.

Substituting an x value into a function produces a corresponding y value. The coordinates of the point (x,y), where y is the largest of all the y values, is said to be a maximum point. The coordinates of the point (x,y), where y is the smallest of all the y values, is said to be a minimum point. To find these points, only a few x values must be tested. First, find all of the x values that make the derivative either zero or undefined. Substitute these values into the original function to obtain the corresponding y values. Compare the y values. The largest y value is a maximum; the smallest y value is a minimum. If the question asks for the maxima or minima on an interval, be certain to also find the y values that correspond to the numbers at either end of the interval.

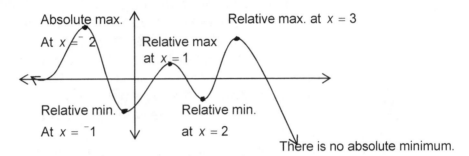

Example:

Find the maxima and minima of $f(x) = 2x^4 - 4x^2$ at the interval $(^-2,1)$.

$f'(x) = 8x^3 - 8x$ Take the derivative first. Find all the x values (critical values) that

$8x^3 - 8x = 0$ make the derivative zero or undefined. In this case, there are no x values that make the

$8x(x^2 - 1) = 0$ derivative undefined.
$8x(x - 1)(x + 1) = 0$ Substitute the critical values into
$x = 0,\ x = 1,\ \text{or}\ x = ^-1$ the original function. Also, plug
$f(0) = 2(0)^4 - 4(0)^2 = 0$ in the endpoint of the interval.
$f(1) = 2(1)^4 - 4(1)^2 = ^-2$ Note that 1 is a critical point
$f(^-1) = 2(^-1)^4 - 4(^-1)^2 = ^-2$ and an endpoint.
$f(^-2) = 2(^-2)^4 - 4(^-2)^2 = 16$

The maximum is at $(^-2,16)$ and there are minima at $(1,^-2)$ and $(^-1,^-2)$. $(0,0)$ is neither the maximum or minimum on $(^-2,1)$ but it is still considered a relative extra point.

The first derivative reveals whether a **curve is rising or falling** (increasing or decreasing) from the left to the right. In much the same way, the second derivative relates whether the curve is concave up or concave down. Curves which are concave up are said to "collect water;" curves which are concave down are said to "dump water." To find the intervals where a curve is concave up or concave down, follow the following steps.

1. Take the second derivative (i.e. the derivative of the first derivative).
2. Find the critical x values.
    -Set the second derivative equal to zero and solve for critical x values.
    -Find the x values that make the second derivative undefined (i.e. make the denominator of the second derivative equal to zero). Such values may not always exist.
3. Pick sample values which are both less than and greater than each of the critical values.
4. Substitute each of these sample values into the second derivative and determine whether the result is positive or negative.
    -If the sample value yields a positive number for the second derivative, the curve is concave up on the interval where the sample value originated.
    -If the sample value yields a negative number for the second derivative, the curve is concave down on the interval where the sample value originated.

Example:

Find the intervals where the curve is concave up and concave down for $f(x) = x^4 - 4x^3 + 16x - 16$.

$f'(x) = 4x^3 - 12x^2 + 16$ Take the second derivative.

$f''(x) = 12x^2 - 24x$ Find the critical values by setting the second derivative equal to zero.

$12x^2 - 24x = 0$
$12x(x - 2) = 0$ There are no values that make the second derivative undefined.

$x = 0$ or $x = 2$

Set up a number line with the critical values.

Sample values: $^-1$, 1, 3 Pick sample values in each of the 3

$f''(^-1) = 12(^-1)^2 - 24(^-1) = 36$ intervals. If the sample value

$f''(1) = 12(1)^2 - 24(1) = ^-12$ produces a negative number,

$f''(3) = 12(3)^2 - 24(3) = 36$ the function is concave down.

If the value produces a positive number, the curve is concave up. If
The value produces a zero, the function is linear.

Therefore when $x < 0$ the function is concave up,
when $0 < x < 2$ the function is concave down,
when $x > 2$ the function is concave up.

A **point of inflection** is a point where a curve changes from being concave up to concave down or vice versa. To find these points, follow the steps for finding the intervals where a curve is concave up or concave down. A critical value is part of an inflection point if the curve is concave up on one side of the value and concave down on the other. The critical value is the x coordinate of the inflection point. To get the y coordinate, plug the critical value into the **original** function.

Example: Find the inflection points of $f(x) = 2x - \tan x$ where $\dfrac{^-\pi}{2} < x < \dfrac{\pi}{2}$.

$(x) = 2x - \tan x \quad \dfrac{^-\pi}{2} < x < \dfrac{\pi}{2}$ Note the restriction on x.

$f'(x) = 2 - \sec^2 x$ Take the second derivative.

$f''(x) = 0 - 2 \bullet \sec x \bullet (\sec x \tan x)$ Use the Power rule.

$= ^-2 \bullet \dfrac{1}{\cos x} \bullet \dfrac{1}{\cos x} \bullet \dfrac{\sin x}{\cos x}$

The derivative of $\sec x$ is $(\sec x \tan x)$.

$f''(x) = \dfrac{^-2\sin x}{\cos^3 x}$

Find critical values by solving for the second derivative equal to zero.

$$0 = \frac{^-2\sin x}{\cos^3 x}$$

No x values on $\left(\dfrac{^-\pi}{2}, \dfrac{\pi}{2}\right)$ make the denominator zero.

$^-2\sin x = 0$

$\sin x = 0$

$x = 0$

Pick sample values on each side of the critical value $x = 0$.

Sample values: $x = \dfrac{^-\pi}{4}$ and $x = \dfrac{\pi}{4}$

$$f''\left(\frac{^-\pi}{4}\right) = \frac{^-2\sin(^-\pi/4)}{\cos^3(\pi/4)} = \frac{^-2(^-\sqrt{2}/2)}{(\sqrt{2}/2)^3} = \frac{\sqrt{2}}{(\sqrt{8}/8)} = \frac{8\sqrt{2}}{\sqrt{8}} = \frac{8\sqrt{2}}{\sqrt{8}} \cdot \frac{\sqrt{8}}{\sqrt{8}}$$

$$= \frac{8\sqrt{16}}{8} = 4$$

$$f''\left(\frac{\pi}{4}\right) = \frac{^-2\sin(\pi/4)}{\cos^3(\pi/4)} = \frac{^-2(\sqrt{2}/2)}{(\sqrt{2}/2)^3} = \frac{^-\sqrt{2}}{(\sqrt{8}/8)} = \frac{^-8\sqrt{2}}{\sqrt{8}} = -4$$

The second derivative is positive on $(0, \infty)$ and negative on $(^-\infty, 0)$. So the curve changes concavity at $x = 0$. Use the original equation to find the y value that inflection occurs at.

$f(0) = 2(0) - \tan 0 = 0 - 0 = 0$

The inflection point is $(0,0)$.

Taking the antiderivative of a function is the opposite of taking the derivative of the function--much in the same way that squaring an expression is the opposite of taking the square root of the expression. For example, since the derivative of x^2 is $2x$ then the antiderivative of 2x is x^2. The key to being able to take antiderivatives is being as familiar as possible with the derivative rules.

To take the antiderivative of an algebraic function (the sum of products of coefficients and variables raised to powers other than negative one), take the antiderivative of each term in the function by following these steps.

1. Take the antiderivative of each term separately.

2. The coefficient of the variable should be equal to one plus the exponent.

3. If the coefficient is not one more than the exponent, put the correct coefficient on the variable and also multiply by the reciprocal of the number put in.

> Ex. For $4x^5$, the coefficient should be 6 not 4. So put in 6 and the reciprocal 1/6 to achieve $(4/6)6x^5$.

4. Finally take the antiderivative by replacing the coefficient and variable with just the variable raised to one plus the original exponent.

> Ex. For $(4/6)6x^5$, the antiderivative is $(4/6)x^6$.

You have to add in constant c because there is no way to know if a constant was originally present since the derivative of a constant is zero.

5. Check your work by taking the first derivative of your answer. You should get the original algebraic function.

Examples: Take the antiderivative of each function.

1. $f(x) = 5x^4 + 2x$ The coefficient of each term is already 1 more than the exponent.

    $F(x) = x^5 + x^2 + c$ $F(x)$ is the antiderivative of $f(x)$

    $F'(x) = 5x^4 + 2x$ Check by taking the derivative of $F(x)$.

2. $f(x) = {}^{-}2x^{-3}$

    $F(x) = x^{-2} + c = \dfrac{1}{x^2} + c$ $F(x)$ is the antiderivative of $f(x)$.

    $F'(x) = {}^{-}2x^{-3}$ Check.

3. $f(x) = {}^{-}4x^2 + 2x^7$ Neither coefficient is correct.

    $f(x) = {}^{-}4 \bullet \dfrac{1}{3} \bullet 3x^2 + 2 \bullet \dfrac{1}{8} \bullet 8x^7$

                            Put in the correct coefficient along with its reciprocal.

    $F(x) = {}^{-}4 \bullet \dfrac{1}{3}x^3 + 2 \bullet \dfrac{1}{8}x^8 + c$ $F(x)$ is the antiderivative of $f(x)$.

    $F(x) = \dfrac{{}^{-}4}{3}x^3 + \dfrac{1}{4}x^8 + c$

    $F'(x) = {}^{-}4x^2 + 2x^7$ Check.

The rules for taking **antiderivatives of trigonometric functions** follow, but be aware that these can get very confusing to memorize because they are very similar to the derivative rules. Check the antiderivative you get by taking the derivative and comparing it to the original function.

1. $\sin x$ the antiderivative for $\sin x$ is $^-\cos x + c$.

2. $\cos x$ the antiderivative for $\cos x$ is $\sin x + c$.

3. $\tan x$ the antiderivative for $\tan x$ is $-\ln|\cos x| + c$.

4. $\sec^2 x$ the antiderivative for $\sec^2 x$ is $\tan x + c$.

5. $\sec x \tan x$ the antiderivative for $\sec x \tan x$ is $\sec x + c$.

If the trigonometric function has a coefficient, simply keep the coefficient and multiply the antiderivative by it.

Examples: Find the antiderivatives for the following functions.

1. $f(x) = 2\sin x$ Carry the 2 throughout the problem.

$\quad F(x) = 2(^-\cos x) = {}^-2\cos x + c$ $F(x)$ is the antiderivative.

$\quad F'(x) = {}^-2(^-\sin x) = 2\sin x$ Check by taking the derivative of $F(x)$.

2. $f(x) = \dfrac{\tan x}{5}$

$\quad F(x) = \dfrac{-\ln|\cos x|}{5} + c$ $F(x)$ is the antiderivative of $f(x)$.

$\quad F'(x) = \dfrac{1}{5}\left(-\dfrac{1}{|\cos x|}\right)(-\sin x) = \dfrac{1}{5}\tan x$ Check by taking the derivative of $F(x)$.

Practice problems: Find the antiderivative of each function.

1. $f(x) = {}^-20\cos x$ 2. $f(x) = \pi \sec x \tan x$

The antiderivative, or indefinite integral, of a given function f is a function F whose derivative is equal to f (i.e. F' = f).
Use the following rules when finding the **antiderivative of an exponential function**.

1. e^x The antiderivative of e^x is the same $e^x + c$.
2. a^x The antiderivative of a^x, where a is any number, is $a^x/\ln a + c$.

Examples: Find the antiderivatives of the following functions:
1. $f(x) = 10e^x$

$F(x) = 10e^x + c$ $F(x)$ is the antiderivative.

$F'(x) = 10e^x$ Check by taking the derivative of $F(x)$.

2. $f(x) = \dfrac{2^x}{3}$

$F(x) = \dfrac{1}{3} \cdot \dfrac{2^x}{\ln 2} + c$ $F(x)$ is the antiderivative.

$F'(x) = \dfrac{1}{3\ln 2} \ln 2(2^x)$ Check by taking the derivative of $F(x)$.

$F'(x) = \dfrac{2^x}{3}$

Taking the integral of a function and evaluating it from one x value to another provides the **total area under the curve** (i.e. between the curve and the x axis). Remember, though, that regions above the x axis have "positive" area and regions below the x axis have "negative" area. You must account for these positive and negative values when finding the area under curves. Follow these steps.

1. Determine the x values that will serve as the left and right boundaries of the region.
2. Find all x values between the boundaries that are either solutions to the function or are values which are not in the domain of the function. These numbers are the interval numbers.
3. Integrate the function.
4. Evaluate the integral once for each of the intervals using the boundary numbers.
5. If any of the intervals evaluates to a negative number, make it positive (the negative simply tells you that the region is below the x axis).
6. Add the value of each integral to arrive at the area under the curve.

Example:

Find the area under the following function on the given intervals.

$f(x) = \sin x$; $(0, 2\pi)$

$\sin x = 0$ Find any roots to f(x) on $(0, 2\pi)$.

$x = \pi$

$(0, \pi)$ $(\pi, 2\pi)$ Determine the intervals using the boundary numbers and the roots.

$\int \sin x\, dx = ^{-}\cos x$ Integrate f(x). We can ignore the constant c because we have numbers to use to evaluate the

$^{-}\cos x\Big]_{x=0}^{x=\pi} = ^{-}\cos \pi - (^{-}\cos 0)$ integral.

$^{-}\cos x\Big]_{x=0}^{x=\pi} = ^{-}(-1) + (1) = 2$

$^{-}\cos x\Big]_{x=\pi}^{x=2\pi} = ^{-}\cos 2\pi - (^{-}\cos \pi)$

$^{-}\cos x\Big]_{x=\pi}^{x=2\pi} = ^{-}1 + (^{-}1) = ^{-}2$ The $^{-}2$ means that for $(\pi, 2\pi)$, the region is below the x axis, but the area is still 2.

Area $= 2 + 2 = 4$ Add the 2 integrals together to get the area.

Finding the **area between two curves** is much the same as finding the area under one curve. But instead of finding the roots of the functions, you need to find the x values which produce the same number from both functions (set the functions equal and solve). Use these numbers and the given boundaries to write the intervals. On each interval you must pick sample values to determine which function is "on top" of the other. Find the integral of each function. For each interval, subtract the "bottom" integral from the "top" integral. Use the interval numbers to evaluate each of these differences. Add the evaluated integrals to get the total area between the curves.

Example:

Find the area of the regions bounded by the two functions on the indicated intervals.

$$f(x) = x + 2 \text{ and } g(x) = x^2 \quad \left[{}^-2, 3 \right]$$

Set the functions equal and solve.

$$x + 2 = x^2$$
$$0 = x^2 - x - 2$$
$$0 = (x - 2)(x + 1)$$
$$x = 2 \text{ or } x = {}^-1$$

Use the solutions and the boundary numbers to write the intervals.

$$({}^-2, {}^-1) \ ({}^-1, 2) \ (2, 3)$$

$$f({}^-3/2) = \left(\frac{{}^-3}{2} \right) + 2 = \frac{1}{2}$$

Pick sample values on the integral and evaluate each function as that number.

$$g({}^-3/2) = \left(\frac{{}^-3}{2} \right)^2 = \frac{9}{4}$$

$g(x)$ is "on top" on $\left[{}^-2, {}^-1 \right]$.

$$f(0) = 2$$

$f(x)$ is "on top" on $\left[{}^-1, 2 \right]$.

$$g(0) = 0$$

$$f(5/2) = \frac{5}{2} + 2 = \frac{9}{2}$$

$g(x)$ is "on top" on $[2, 3]$.

$$g(5/2) = \left(\frac{5}{2} \right)^2 = \frac{25}{4}$$

$$\int f(x)dx = \int (x + 2)dx$$

$$\int f(x)dx = \int xdx + 2\int dx$$

$$\int f(x)dx = \frac{1}{1+1}x^{1+1} + 2x$$

$$\int f(x)dx = \frac{1}{2}x^2 + 2x$$

$$\int g(x)dx = \int x^2 dx$$

$$\int g(x)dx = \frac{1}{2+1}x^{2+1} = \frac{1}{3}x^3$$

$$\text{Area } 1 = \int g(x)dx - \int f(x)dx$$

g(x) is "on top" on $\left[^-2, ^-1\right]$.

$$\text{Area } 1 = \frac{1}{3}x^3 - \left(\frac{1}{2}x^2 + 2x\right)\Big]_{-2}^{-1}$$

Area

$$1 = \left[\frac{1}{3}(^-1)^3 - \left(\frac{1}{2}(^-1)^2 + 2(^-1)\right)\right] - \left[\frac{1}{3}(^-2)^3 - \left(\frac{1}{2}(^-2)^2 + 2(^-2)\right)\right]$$

$$\text{Area } 1 = \left[\frac{^-1}{3} - \left(\frac{^-3}{2}\right)\right] - \left[\frac{^-8}{3} - (^-2)\right]$$

$$\text{Area } 1 = \left(\frac{7}{6}\right) - \left(\frac{^-2}{3}\right) = \frac{11}{6}$$

$$\text{Area } 2 = \int f(x)dx - \int g(x)dx$$

f(x) is "on top" on $\left[^-1, 2\right]$.

$$\text{Area } 2 = \frac{1}{2}x^2 + 2x - \frac{1}{3}x^3\Big]_{-1}^{2}$$

$$\text{Area } 2 = \left(\frac{1}{2}(2)^2 + 2(2) - \frac{1}{3}(2)^3\right) - \left(\frac{1}{2}(^-1)^2 + 2(^-1) - \frac{1}{3}(^-1)^3\right)$$

$$\text{Area } 2 = \left(\frac{10}{3}\right) - \left(\frac{1}{2} - 2 + \frac{1}{3}\right)$$

$$\text{Area } 2 = \frac{27}{6}$$

$$\text{Area } 3 = \int g(x)dx - \int f(x)dx$$

g(x) is "on top" on [2,3].

$$\text{Area } 3 = \frac{1}{3}x^3 - \left(\frac{1}{2}x^2 + 2x\right)\Big]_{2}^{3}$$

$$\text{Area } 3 = \left[\frac{1}{3}(3)^3 - \left(\frac{1}{2}(3^2) + 2(3)\right)\right] - \left[\frac{1}{3}(2)^3 - \left(\frac{1}{2}(2)^2 + 2(2)\right)\right]$$

$$\text{Area } 3 = \left(\frac{^-3}{2}\right) - \left(\frac{^-10}{3}\right) = \frac{11}{6}$$

$$\text{Total area} = \frac{11}{6} + \frac{27}{6} + \frac{11}{6} = \frac{49}{6} = 8\frac{1}{6}$$

If you take the area bounded by a curve or curves and revolve it about a line, the result is a solid of revolution. To find the volume of such a solid, the Washer Method works in most instances. Imagine slicing through the solid perpendicular to the line of revolution. The "slice" should resemble a washer. Use an integral and the formula for the volume of disk.

$$Volume_{disk} = \pi \bullet radius^2 \bullet thickness$$

Depending on the situation, the radius is the distance from the line of revolution to the curve; or if there are two curves involved, the radius is the difference between the two functions. The thickness is dx if the line of revolution is parallel to the x axis and dy if the line of revolution is parallel to the y axis. Finally, integrate the volume expression using the boundary numbers from the interval.

Example:

Find the value of the solid of revolution found by revolving $f(x) = 9 - x^2$ about the x axis on the interval $[0, 4]$.

$radius = 9 - x^2$
$thickness = dx$

$Volume = \int_0^4 \pi(9 - x^2)^2 dx$ Use the formula for volume of a disk.

$Volume = \pi \int_0^4 \left(81 - 18x^2 + x^4\right) dx$

$Volume = \pi \left(81x - \dfrac{18}{2+1}x^3 + \dfrac{1}{4+1}x^5 \right)]_0^4$ Take the integral.

$Volume = \pi \left(81x - 6x^3 + \dfrac{1}{5}x^5 \right)]_0^4$ Evaluate the integral first $x = 4$ then at $x = 0$

$Volume = \pi \left[\left(324 - 384 + \dfrac{1024}{5} \right) - \left(0 - 0 + 0 \right) \right]$

$Volume = \pi \left(144\dfrac{4}{5} \right) = 144\dfrac{4}{5}\pi = 454.9$

SUBAREA IV–MEASUREMENT AND GEOMETRY

0014. Understand and apply measurement principles.

There are many methods for converting measurements within a system. One method is to multiply the given measurement by a conversion factor. This conversion factor is the ratio of:

$$\frac{\text{new units}}{\text{old units}} \quad \text{OR} \quad \frac{\text{what you want}}{\text{what you have}}$$

Sample problems:

1. Convert 3 miles to yards.

$$\frac{3 \text{ miles}}{1} \times \frac{1{,}760 \text{ yards}}{1 \text{ mile}} = \frac{\text{yards}}{}$$

1. multiply by the conversion factor
2. cancel the miles units
3. solve

$$= 5{,}280 \text{ yards}$$

2. Convert 8,750 meters to kilometers.

$$\frac{8{,}750 \text{ meters}}{1} \times \frac{1 \text{ kilometer}}{1000 \text{ meters}} = \frac{\text{km}}{}$$

1. multiply by the conversion factor
2. cancel the meters units
3. solve

$$= 8.75 \text{ kilometers}$$

Use the formulas to find the volume and surface area.

FIGURE	VOLUME	TOTAL SURFACE AREA
Right Cylinder	$\pi r^2 h$	$2\pi rh + 2\pi r^2$
Right Cone	$\dfrac{\pi r^2 h}{3}$	$\pi r\sqrt{r^2 + h^2} + \pi r^2$
Sphere	$\dfrac{4}{3}\pi r^3$	$4\pi r^2$
Rectangular Solid	LWH	$2LW + 2WH + 2LH$

Note: $\sqrt{r^2 + h^2}$ is equal to the slant height of the cone.

Sample problem:

1. Given the figure below, find the volume and surface area.

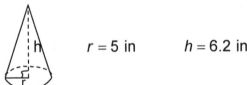

$r = 5$ in $h = 6.2$ in

Volume = $\dfrac{\pi r^2 h}{3}$ First write the formula.

$\dfrac{1}{3}\pi(5^2)(6.2)$ Then substitute.

162.3 cubic inches   Finally solve the problem.

Surface area = $\pi r\sqrt{r^2 + h^2} + \pi r^2$ First write the formula.

$\pi 5\sqrt{5^2 + 6.2^2} + \pi 5^2$ Then substitute.

203.6 square inches   Compute.

Note: volume is always given in cubic units and area is always given in square units.

FIGURE	AREA FORMULA	PERIMETER FORMULA
Rectangle	LW	$2(L+W)$
Triangle	$\frac{1}{2}bh$	$a+b+c$
Parallelogram	bh	sum of lengths of sides
Trapezoid	$\frac{1}{2}h(a+b)$	sum of lengths of sides

Sample problems:

1. Find the area and perimeter of a rectangle if its length is 12 inches and its diagonal is 15 inches.

1. Draw and label sketch.

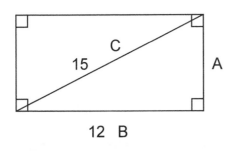

2. Since the height is still needed use Pythagorean formula to find missing leg of the triangle.

$$A^2 + B^2 = C^2$$
$$A^2 + 12^2 = 15^2$$
$$A^2 = 15^2 - 12^2$$
$$A^2 = 81$$
$$A = 9$$

Now use this information to find the area and perimeter.

$A = LW$	$P = 2(L+W)$	1. write formula
$A = (12)(9)$	$P = 2(12+9)$	2. substitute
$A = 108 \text{ in}^2$	$P = 42$ inches	3. solve

Given a circular figure the formulas are as follows:

$$A = \pi r^2 \qquad\qquad C = \pi d \quad \text{or} \quad 2\pi r$$

Sample problem:

1. If the area of a circle is 50 cm^2, find the circumference.

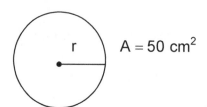

$A = 50$ cm^2

1. Draw sketch.

2. Determine what is still needed.

Use the area formula to find the radius.

$A = \pi r^2$	1. write formula
$50 = \pi r^2$	2. substitute
$\dfrac{50}{\pi} = r^2$	3. divide by π
$15.915 = r^2$	4. substitute
$\sqrt{15.915} = \sqrt{r^2}$	5. take square root of both sides
$3.989 \approx r$	6. compute

Use the approximate answer (due to rounding) to find the circumference.

$C = 2\pi r$	1. write formula
$C = 2\pi (3.989)$	2. substitute
$C \approx 25.064$	3. compute

When using formulas to find each of the required items it is helpful to remember to always use the same strategies for problem solving. First, draw and label a sketch if needed. Second, write the formula down and then substitute in the known values. This will assist in identifying what is still needed (the unknown). Finally, solve the resulting equation.

Being consistent in the strategic approach to problem solving is paramount to teaching the concept as well as solving it.

Use appropriate problem solving strategies to find the solution.

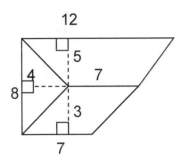

1. Find the area of the given figure.

2. Cut the figure into familiar shapes.

3. Identify what type figures are given and write the appropriate formulas.

Area of figure 1 (triangle)	Area of figure 2 (parallelogram)	Area of figure 3 (trapezoid)
$A = \frac{1}{2}bh$	$A = bh$	$A = \frac{1}{2}h(a+b)$
$A = \frac{1}{2}(8)(4)$	$A = (7)(3)$	$A = \frac{1}{2}(5)(12+7)$
$A = 16$ sq. ft	$A = 21$ sq. ft	$A = 47.5$ sq. ft

Now find the total area by adding the area of all figures.

Total area $= 16 + 21 + 47.5$
Total area $= 84.5$ square ft

Given the figure below, find the area by dividing the polygon into smaller shapes.

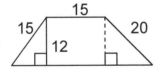

1. divide the figure into two triangles and a rectangle.

2. find the missing lengths.

3. find the area of each part.

4. find the sum of all areas.

Find base of both right triangles using Pythagorean Formula:

$$a^2 + b^2 = c^2$$
$$a^2 + 12^2 = 15^2$$
$$a^2 = 225 - 144$$
$$a^2 = 81$$
$$a = 9$$

$$a^2 + b^2 = c^2$$
$$a^2 + 12^2 = 20^2$$
$$a^2 = 400 - 144$$
$$a^2 = 256$$
$$a = 16$$

Area of triangle 1   Area of triangle 2   Area of rectangle

$$A = \frac{1}{2}bh \qquad\qquad A = \frac{1}{2}bh \qquad\qquad A = LW$$

$$A = \frac{1}{2}(9)(12) \qquad A = \frac{1}{2}(16)(12) \qquad A = (15)(12)$$

A = 54 sq. units   A = 96 sq. units   A = 180 sq. units

Find the sum of all three figures.

$$54 + 96 + 180 = 330 \text{ square units}$$

Polygons are similar if and only if there is a one-to-one correspondence between their vertices such that the corresponding angles are congruent and the lengths of corresponding sides are proportional.

Given the rectangles below, compare the area and perimeter.

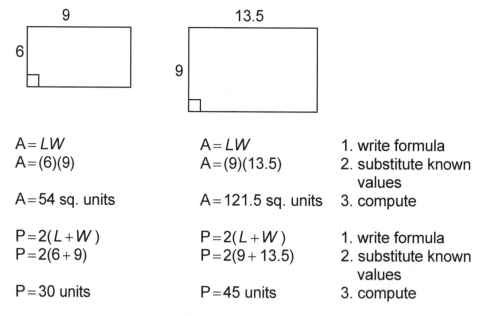

$A = LW$ $A = LW$ 1. write formula
$A = (6)(9)$ $A = (9)(13.5)$ 2. substitute known
                                        values
$A = 54$ sq. units $A = 121.5$ sq. units   3. compute

$P = 2(L + W)$ $P = 2(L + W)$ 1. write formula
$P = 2(6 + 9)$ $P = 2(9 + 13.5)$ 2. substitute known
                                        values
$P = 30$ units $P = 45$ units   3. compute

Notice that the areas relate to each other in the following manner:

Ratio of sides $9/13.5 = 2/3$

Multiply the first area by the square of the reciprocal $(3/2)^2$ to get the second area.

$$54 \times (3/2)^2 = 121.5$$

The perimeters relate to each other in the following manner:

Ratio of sides $9/13.5 = 2/3$

Multiply the perimeter of the first by the reciprocal of the ratio to get the perimeter of the second.

$$30 \times 3/2 = 45$$

FIGURE	LATERAL AREA	TOTAL AREA	VOLUME
Right prism	sum of area of lateral faces (rectangles)	lateral area plus 2 times the area of base	area of base times height
regular pyramid	sum of area of lateral faces (triangles)	lateral area plus area of base	1/3 times the area of the base times the height

Find the total area of the given figure:

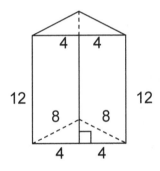

1. Since this is a triangular prism, first find the area of the bases.

2. Find the area of each rectangular lateral face.

3. Add the areas together.

$A = \dfrac{1}{2}bh$ $\qquad$ $A = LW$ $\qquad$ 1. write formula

$8^2 = 4^2 + h^2$ $\qquad\qquad$ 2. find the height of

$h = 6.928$ $\qquad\qquad\qquad$ the base triangle

$A = \dfrac{1}{2}(8)(6.928)$ $\quad$ $A = (8)(12)$ $\qquad$ 3. substitute known values

$A = 27.713$ sq. units $\;$ $A = 96$ sq. units $\qquad$ 4. compute

Total Area $= 2(27.713) + 3(96)$
$\qquad\qquad = 343.426$ sq. units

FIGURE	VOLUME	TOTAL SURFACE AREA	LATERAL AREA
Right Cylinder	$\pi r^2 h$	$2\pi rh + 2\pi r^2$	$2\pi rh$
Right Cone	$\dfrac{\pi r^2 h}{3}$	$\pi r\sqrt{r^2 + h^2} + \pi r^2$	$\pi r\sqrt{r^2 + h^2}$

Note: $\sqrt{r^2 + h^2}$ is equal to the slant height of the cone.

Sample problem:

1. A water company is trying to decide whether to use traditional cylindrical paper cups or to offer conical paper cups since both cost the same. The traditional cups are 8 cm wide and 14 cm high. The conical cups are 12 cm wide and 19 cm high. The company will use the cup that holds the most water.

1. Draw and label a sketch of each.

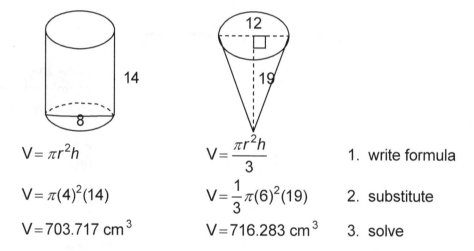

$V = \pi r^2 h$	$V = \dfrac{\pi r^2 h}{3}$	1. write formula
$V = \pi(4)^2(14)$	$V = \dfrac{1}{3}\pi(6)^2(19)$	2. substitute
$V = 703.717$ cm^3	$V = 716.283$ cm^3	3. solve

The choice should be the conical cup since its volume is more.

FIGURE	VOLUME	TOTAL SURFACE AREA
Sphere	$\dfrac{4}{3}\pi r^3$	$4\pi r^2$

Sample problem:

1. How much material is needed to make a basketball that has a diameter of 15 inches? How much air is needed to fill the basketball?

Draw and label a sketch:

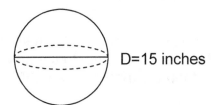 D=15 inches

Total surface area   Volume

$TSA = 4\pi r^2$ $V = \dfrac{4}{3}\pi r^3$ 1. write formula

$\quad = 4\pi(7.5)^2$ $\quad = \dfrac{4}{3}\pi(7.5)^3$ 2. substitute

$\quad = 706.9 \text{ in}^2$ $\quad = 1767.1 \text{ in}^3$ 3. solve

Pythagorean theorem states that the square of the length of the hypotenuse is equal to the sum of the squares of the lengths of the legs. Symbolically, this is stated as:

$$c^2 = a^2 + b^2$$

Given the right triangle below, find the missing side.

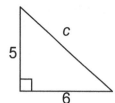

$c^2 = a^2 + b^2$ 1. write formula

$c^2 = 5^2 + 6^2$ 2. substitute known values

$c^2 = 61$ 3. take square root

$c = \sqrt{61}$ or 7.81   4. solve

The Converse of the Pythagorean Theorem states that if the square of one side of a triangle is equal to the sum of the squares of the other two sides, then the triangle is a right triangle.

Example:
Given ΔXYZ, with sides measuring 12, 16 and 20 cm. Is this a right triangle?

$$c^2 = a^2 + b^2$$
$$20^2 \ ? \ 12^2 + 16^2$$
$$400 \ ? \ 144 + 256$$
$$400 = 400$$

Yes, the triangle is a right triangle.

This theorem can be expanded to determine if triangles are obtuse or acute.
If the square of the longest side of a triangle is greater than the sum of the squares of the other two sides, then the triangle is an obtuse triangle.
and
If the square of the longest side of a triangle is less than the sum of the squares of the other two sides, then the triangle is an acute triangle.

Example:
Given ΔLMN with sides measuring 7, 12, and 14 inches. Is the triangle right, acute, or obtuse?

$$14^2 \ ? \ 7^2 + 12^2$$
$$196 \ ? \ 49 + 144$$
$$196 > 193 \qquad \text{Therefore, the triangle is obtuse.}$$

0015. Understand the principles and properties of axiomatic (synthetic) geometries.

The classifying of angles refers to the angle measure. The naming of angles refers to the letters or numbers used to label the angle.

Sample Problem:

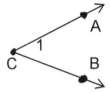

$\overrightarrow{CA}$ (read ray CA) and $\overrightarrow{CB}$ are the sides of the angle.
The angle can be called $\angle ACB$, $\angle BCA$, $\angle C$ or $\angle 1$.

Angles are classified according to their size as follows:

acute:	greater than 0 and less than 90 degrees.
right:	exactly 90 degrees.
obtuse:	greater than 90 and less than 180 degrees.
straight:	exactly 180 degrees

Angles can be classified in a number of ways. Some of those classifications are outlined here.

Adjacent angles have a common vertex and one common side but no interior points in common.

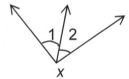

Complimentary angles add up to 90 degrees.

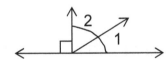

Supplementary angles add up to 180 degrees.

Vertical angles have sides that form two pairs of opposite rays.

Corresponding angles are in the same corresponding position on two parallel lines cut by a transversal.

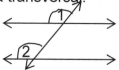

Alternate interior angles are diagonal angles on the inside of two parallel lines cut by a transversal.

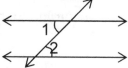

Alternate exterior angles are diagonal angles on the outside of two parallel lines cut by a transversal.

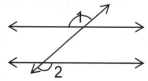

Parallel lines or planes do not intersect.

Perpendicular lines or planes form a 90 degree angle to each other.

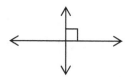

Intersecting lines share a common point and intersecting planes share a common set of points or line.

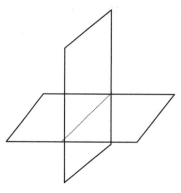

Skew lines do not intersect and do not lie on the same plane.

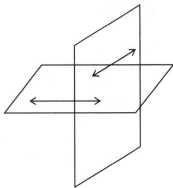

Two triangles are congruent if each of the three angles and three sides of one triangle match up in a one-to-one fashion with congruent angles and sides of the second triangle. In order to see how the sides and angles match up, it is sometimes necessary to imagine rotating or reflecting one of the triangles so the two figures are oriented in the same position.

There are shortcuts to the above procedure for proving two triangles congruent.

Side-Side-Side (SSS) Congruence--If the three sides of one triangle match up in a one-to-one congruent fashion with the three sides of the other triangle, then the two triangles are congruent. With SSS it is not necessary to even compare the angles; they will automatically be congruent.

Angle-Side-Angle (ASA) Congruence--If two angles of one triangle match up in a one-to-one congruent fashion with two angles in the other triangle and if the sides between the two angles are also congruent, then the two triangles are congruent. With ASA the sides that are used for congruence must be located between the two angles used in the first part of the proof.

Side-Angle-Side (SAS) Congruence--If two sides of one triangle match up in a one-to-one congruent fashion with two sides in the other triangle and if the angles between the two sides are also congruent, then the two triangles are congruent. With SAS the angles that are used for congruence must be located between the two sides used in the first part of the proof.

In addition to SSS, ASA, and SAS, **Angle-Angle-Side (AAS)** is also a congruence shortcut.

AAS states that if two angles of one triangle match up in a one-to-one congruent fashion with two angles in the other triangle and if two sides that are not between the aforementioned sets of angles are also congruent, then the triangles are congruent. ASA and AAS are very similar; the only difference is where the congruent sides are located. If the sides are between the congruent sets of angles, use ASA. If the sides are not located between the congruent sets of angles, use AAS.

Hypotenuse-Leg (HL) is a congruence shortcut which can only be used with right triangles. If the hypotenuse and leg of one right triangle are congruent to the hypotenuse and leg of the other right triangle, then the two triangles are congruent.

Two triangles are overlapping if a portion of the interior region of one triangle is shared in common with all or a part of the interior region of the second triangle.

The most effective method for proving two overlapping triangles congruent is to draw the two triangles separated. Separate the two triangles and label all of the vertices using the labels from the original overlapping figures. Once the separation is complete, apply one of the congruence shortcuts: SSS, ASA, SAS, AAS, or HL.

A parallelogram is a quadrilateral (four-sided figure) in which opposite sides are parallel. There are three shortcuts for proving that a quadrilateral is a parallelogram without directly showing that the opposite sides are parallel.

If the diagonals of a quadrilateral bisect each other, then the quadrilateral is also a parallelogram. Note that this shortcut only requires the diagonals to bisect each other; the diagonals do not need to be congruent.

If both pairs of opposite sides are congruent, then the quadrilateral is a parallelogram.

If both pairs of opposite angles are congruent, then the quadrilateral is a parallelogram.

If one pair of opposite sides are both parallel and congruent, then the quadrilateral is a parallelogram.

The following table illustrates the properties of each quadrilateral.

	Parallel Opposite Sides	Bisecting Diagonals	Equal Opposite Sides	Equal Opposite Angles	Equal Diagonals	All Sides Equal	All Angles Equal	Perpen-dicular Diagonals
Parallelogram	X	X	X	X				
Rectangle	X	X	X	X	X		X	
Rhombus	X	X	X	X		X		X
Square	X	X	X	X	X	X	X	X

A trapezoid is a quadrilateral with exactly one pair of parallel sides. A trapezoid is different from a parallelogram because a parallelogram has two pairs of parallel sides.

The two parallel sides of a trapezoid are called the bases, and the two non-parallel sides are called the legs. If the two legs are the same length, then the trapezoid is called isosceles.

The segment connecting the two midpoints of the legs is called the median. The median has the following two properties.

The median is parallel to the two bases.

The length of the median is equal to one-half the sum of the length of the two bases.

The segment joining the midpoints of two sides of a triangle is called a **median**. All triangles have three medians. Each median has the following two properties.

A median is parallel to the third side of the triangle.
The length of a median is one-half the length of the third side of the triangle.

Every **angle** has exactly one ray which bisects the angle. If a point on such a bisector is located, then the point is equidistant from the two sides of the angle. Distance from a point to a side is measured along a segment which is perpendicular to the angle's side. The converse is also true. If a point is equidistant from the sides of an angle, then the point is on the bisector of the angle.

Every **segment** has exactly one line which is both perpendicular to and bisects the segment. If a point on such a perpendicular bisector is located, then the point is equidistant to the endpoints of the segment. The converse is also true. If a point is equidistant from the endpoints of a segment, then that point is on the perpendicular bisector of the segment.

Every segment has exactly one line which is both perpendicular to and bisects the segment. If a point on such a perpendicular bisector is located, then the point is equidistant to the endpoints of the segment. The converse is also true. If a point is equidistant from the endpoints of a segments, then that point is on the perpendicular bisector of the segment.

A median is a segment that connects a vertex to the midpoint of the side opposite from that vertex. Every triangle has exactly three medians.

An altitude is a segment which extends from one vertex and is perpendicular to the side opposite that vertex. In some cases, the side opposite from the vertex used will need to be extended in order for the altitude to form a perpendicular to the opposite side. The length of the altitude is used when referring to the height of the triangle.

If the three segments which bisect the three angles of a triangle are drawn, the segments will all intersect in a single point. This point is equidistant from all three sides of the triangle. Recall that the distance from a point to a side is measured along the perpendicular from the point to the side.

If two planes are parallel and a third plane intersects the first two, then the three planes will intersect in two lines which are also parallel.

Given a line and a point which is not on the line but is in the same plane, then there is exactly one line through the point which is parallel to the given line and exactly one line through the point which is perpendicular to the given line.

If three or more segments intersect in a single point, the point is called a **point of concurrency**.

The following sets of special segments all intersect in points of concurrency.

1. Angle Bisectors
2. Medians
3. Altitudes
4. Perpendicular Bisectors

The points of concurrency can lie inside the triangle, outside the triangle, or on one of the sides of the triangle. The following table summarizes this information.

Possible Location(s) of the
Points of Concurrency

	Inside the Triangle	Outside the Triangle	On the Triangle
Angle Bisectors	x		
Medians	x		
Altitudes	x	X	x
Perpendicular Bisectors	x	X	x

A circle is inscribed in a triangle if the three sides of the triangle are each tangent to the circle. The center of an inscribed circle is called the incenter of the triangle. To find the incenter, draw the three angle bisectors of the triangle. The point of concurrency of the angle bisectors is the incenter or center of the inscribed circle. Each triangle has only one inscribed circle.

A circle is circumscribed about a triangle if the three vertices of the triangle are all located on the circle. The center of a circumscribed circle is called the circumcenter of the triangle. To find the circumcenter, draw the three perpendicular bisectors of the sides of the triangle. The point of concurrency of the perpendicular bisectors is the circumcenter or the center of the circumscribing circle. Each triangle has only one circumscribing circle.

A median is a segment which connects a vertex to the midpoint of the side opposite that vertex. Every triangle has three medians. The point of concurrency of the three medians is called the **centroid**.

The centroid divides each median into two segments whose lengths are always in the ratio of 1:2. The distance from the vertex to the centroid is always twice the distance from the centroid to the midpoint of the side opposite the vertex.

If two circles have radii which are in a ratio of $a : b$, then the following ratios are also true for the circles.

The diameters are also in the ratio of $a : b$.
The circumferences are also in the ratio $a : b$.

The areas are in the ratio $a^2 : b^2$, or the ratio of the areas is the square of the ratios of the radii.

Construct the perpendicular bisector of a line segment of a given line segment.

Given a line segment with two endpoints such as A and B, follow these steps to construct the line which both bisects and is perpendicular to the line given segment.

1. Swing an arc of any radius from point A. Swing another arc of the same radius from B. The arcs will intersect at two points. Label these points C and D.

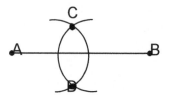

2. Connect C and D to form the perpendicular bisector of segment AB

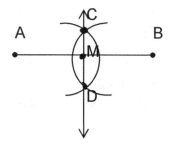

3. The point M where line $\overline{CD}$ and segment $\overline{AB}$ intersect is the midpoint of segment $\overline{AB}$.

Euclid wrote a set of 13 books around 330 B.C. called the Elements. He outlined ten axioms and then deduced 465 theorems. Euclidean geometry is based on the undefined concept of the point, line and plane.

The fifth of Euclid's axioms (referred to as the parallel postulate) was not as readily accepted as the other nine axioms. Many mathematicians throughout the years have attempted to prove that this axiom is not necessary because it could be proved by the other nine. Among the many who attempted to prove this was Carl Friedrich Gauss. His works led to the development of hyperbolic geometry. Elliptical or Reimannian geometry was suggested by G.F. Berhard Riemann. He based his work on the theory of surfaces and used models as physical interpretations of the undefined terms that satisfy the axioms.

The chart below lists the fifth axiom (parallel postulate) as it is given in each of the three geometries.

EUCLIDEAN	ELLIPTICAL	HYPERBOLIC
Given a line and a point not on that line, one and only one line can be drawn through the given point parallel to the given line.	Given a line and a point not on that line, no line can be drawn through the given point parallel to the given line.	Given a line and a point not on that line, two or more lines can be drawn through the point parallel to the given line.

0016. Understand the principles and properties of coordinate geometry.

In order to accomplish the task of finding the distance from a given point to another given line the perpendicular line that intersects the point and line must be drawn and the equation of the other line written. From this information the point of intersection can be found. This point and the original point are used in the distance formula given below:

$$D = \sqrt{(x_2 - x_1)^2 + (y_2 - y_1)^2}$$

Sample Problem:

1. Given the point ($^-4$,3) and the line $y = 4x + 2$, find the distance from the point to the line.

$y = 4x + 2$	1. Find the slope of the given line by solving for y.
$y = 4x + 2$	2. The slope is $4/1$, the perpendicular line will have a slope of $^-1/4$.
$y = \left(^-1/4\right)x + b$	3. Use the new slope and the given point to find the equation of the perpendicular line.
$3 = \left(^-1/4\right)\left(^-4\right) + b$	4. Substitute ($^-4$,3) into the equation.
$3 = 1 + b$	5. Solve.
$2 = b$	6. Given the value for b, write the equation of the perpendicular line.
$y = \left(^-1/4\right)x + 2$	7. Write in standard form.
$x + 4y = 8$	8. Use both equations to solve by elimination to get the point of intersection.

$$^-4x + y = 2$$
$$\underline{x + 4y = 8}$$

9. Multiply the bottom row by 4.

$$^-4x + y = 2$$
$$\underline{4x + 16y = 32}$$
$$17y = 34$$
$$y = 2$$

10. Solve.

$$y = 4x + 2$$
$$2 = 4x + 2$$
$$x = 0$$

11. Substitute to find the x value.
12. Solve.

(0,2) is the point of intersection. Use this point on the original line and the original point to calculate the distance between them.

$$D = \sqrt{(x_2 - x_1)^2 + (y_2 - y_1)^2} \quad \text{where points are (0,2) and (-4,3).}$$

$$D = \sqrt{(^-4 - 0)^2 + (3 - 2)^2} \qquad \text{1. Substitute.}$$

$$D = \sqrt{(16) + (1)} \qquad \text{2. Simplify.}$$

$$D = \sqrt{17}$$

The **distance between two parallel lines**, such as line AB and line CD as shown below is the line segment RS, the perpendicular between the two parallels.

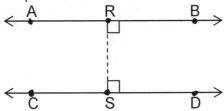

Sample Problem:

Given the geometric figure below, find the distance between the two parallel sides AB and CD.

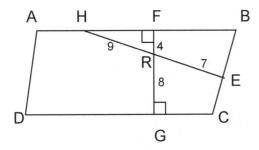

The distance FG is 12 units.
The key to applying the distance formula is to understand the problem before beginning.

$$D = \sqrt{(x_2 - x_1)^2 + (y_2 - y_1)^2}$$

Sample Problem:

1. Find the perimeter of a figure with vertices at $(4,5)$, $(^-4,6)$ and $(^-5,^-8)$.

The figure being described is a triangle. Therefore, the distance for all three sides must be found. Carefully, identify all three sides before beginning.

Side $1 = (4,5)$ to $(^-4,6)$

Side $2 = (^-4,6)$ to $(^-5,^-8)$

Side $3 = (^-5,^-8)$ to $(4,5)$

$$D_1 = \sqrt{(^-4-4)^2 + (6-5)^2} = \sqrt{65}$$

$$D_2 = \sqrt{((^-5-(^-4))^2 + (^-8-6)^2} = \sqrt{197}$$

$$D_3 = \sqrt{((4-(^-5))^2 + (5-(^-8))^2} = \sqrt{250} \text{ or } 5\sqrt{10}$$

$$\text{Perimeter} = \sqrt{65} + \sqrt{197} + 5\sqrt{10}$$

Midpoint Definition:

If a line segment has endpoints of (x_1, y_1) and (x_2, y_2), then the midpoint can be found using:

$$\left(\frac{x_1 + x_2}{2}, \frac{y_1 + y_2}{2} \right)$$

Sample problems:

1. Find the center of a circle with a diameter whose endpoints are $(3,7)$ and $(^-4, ^-5)$.

$$\text{Midpoint} = \left(\frac{3 + (^-4)}{2}, \frac{7 + (^-5)}{2} \right)$$

$$\text{Midpoint} = \left(\frac{^-1}{2}, 1 \right)$$

2. Find the midpoint given the two points $\left(5, 8\sqrt{6}\right)$ and $\left(9, ^-4\sqrt{6}\right)$.

$$\text{Midpoint} = \left(\frac{5 + 9}{2}, \frac{8\sqrt{6} + (^-4\sqrt{6})}{2} \right)$$

$$\text{Midpoint} = \left(7, 2\sqrt{6} \right)$$

Conic sections result from the intersection of a cone and a plane. The three main types of conics are parabolas, ellipses, and hyperbolas.

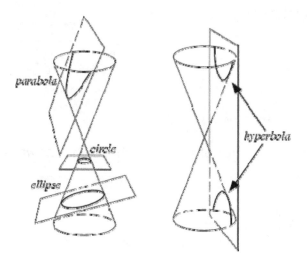

The general equation for a conic section is:

$$Ax^2 + Bxy + Cy^2 + Dx + Ey + F = 0$$

The value of $B^2 - 4AC$ determines the type of conic. If $B^2 - 4AC$ is less than zero the curve is an ellipse or a circle. If equal to zero, the curve is a parabola. If greater than zero, the curve is a hyperbola.

PARABOLAS-A parabola is a set of all points in a plane that are equidistant from a fixed point (focus) and a line (directrix).

FORM OF EQUATION $\quad y = a(x - h)^2 + k \qquad x = a(y - k)^2 + h$

IDENTIFICATION $\;x^2$ term, y not squared $\;\;y^2$ term, x not squared

SKETCH OF GRAPH

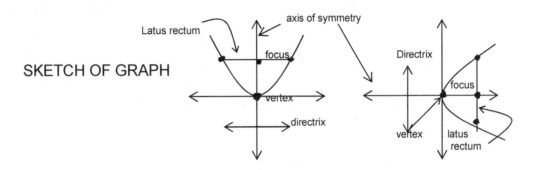

AXIS OF SYMMETRY $\qquad x = h \qquad\qquad\qquad y = k$
   -A line through the vertex and focus upon which the parabola is symmetric.

VERTEX $\qquad\qquad\qquad (h,k) \qquad\qquad\qquad (h,k)$
   -The point where the parabola intersects the axis of symmetry.

FOCUS $\qquad\qquad (h, k + 1/4a) \qquad\qquad (h + 1/4a, k)$

DIRECTRIX $\qquad\qquad y = k - 1/4a \qquad\qquad x = h - 1/4a$

DIRECTION $\qquad\qquad$ up if $a > 0$, $\qquad\qquad$ right if $a > 0$,
OF OPENING $\qquad\qquad$ down if $a < 0$ $\qquad\qquad$ left if $a < 0$

LENGTH OF LATUS $\qquad |1/a| \qquad\qquad\qquad |1/a|$
RECTUM

   -A chord through the focus, perpendicular to the axis of symmetry, with endpoints on the parabola.

Sample Problem:

1. Find all identifying features of $y = {}^-3x^2 + 6x - 1$.

First, the equation must be put into the general form
$y = a(x - h)^2 + k$.

$y = {}^-3x^2 + 6x - 1$ 1. Begin by completing the square.

$= {}^-3(x^2 - 2x + 1) - 1 + 3$

$= {}^-3(x - 1)^2 + 2$ 2. Using the general form of the equation begin to identify known variables.

$a = {}^-3 \quad h = 1 \quad k = 2$

axis of symmetry: $x = 1$
vertex: $(1, 2)$
focus: $\left(1, \, 1\frac{1}{4}\right)$
directrix: $y = 2\frac{3}{4}$
direction of opening: down since $a < 0$
length of latus rectum: $1/3$

ELLIPSE

FORM OF
EQUATION

$$\frac{(x-h)^2}{a^2}+\frac{(y-k)^2}{b^2}=1 \qquad \frac{(x-h)^2}{b^2}+\frac{(y-k)^2}{a^2}=1$$

(for ellipses where   where $b^2=a^2-c^2$ where $b^2=a^2-c^2$
$a^2>b^2$).

IDENTIFICATION   horizontal major axis   vertical major axis

SKETCH

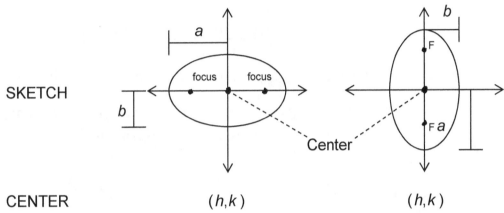

Center

CENTER (h,k) (h,k)

FOCI $(h\pm c,k)$ $(h,k\pm c)$

MAJOR AXIS LENGTH $2a$ $2a$

MINOR AXIS LENGTH $2b$ $2b$

Sample Problem:

Find all identifying features of the ellipse $2x^2+y^2-4x+8y-6=0$.

First, begin by writing the equation in standard form for an ellipse.

$2x^2+y^2-4x+8y-6=0$ 1. Complete the square for
                                           each variable.

$2(x^2-2x+1)+(y^2+8y+16)=6+2(1)+16$

$2(x-1)^2+(y+4)^2=24$ 2. Divide both sides by 24.

$\dfrac{(x-1)^2}{12}+\dfrac{(y+4)^2}{24}=1$ 3. Now the equation is in
                                           standard form.

Identify known variables: $h = 1$ $k = {}^-4$ $a = \sqrt{24}$ or $2\sqrt{6}$

$b = \sqrt{12}$ or $2\sqrt{3}$ $c = 2\sqrt{3}$

Identification: vertical major axis

Center: $(1, {}^-4)$

Foci: $(1, {}^-4 \pm 2\sqrt{3})$

Major axis: $4\sqrt{6}$

Minor axis: $4\sqrt{3}$

HYPERBOLA

FORM OF EQUATION	$\dfrac{(x-h)^2}{a^2} - \dfrac{(y-k)^2}{b^2} = 1$ where $c^2 = a^2 + b^2$	$\dfrac{(y-k)^2}{a^2} - \dfrac{(x-h)^2}{b^2} = 1$ where $c^2 = a^2 + b^2$
IDENTIFICATION	horizontal transverse axis (y^2 is negative)	vertical transverse axis (x^2 is negative)
SKETCH	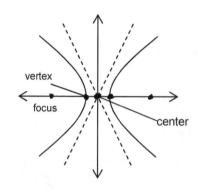	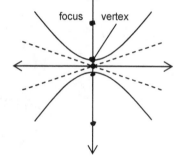
SLOPE OF ASYMPTOTES	$\pm(b/a)$	$\pm(b/a)$
TRANSVERSE AXIS (endpoints are vertices of the hyperbola and go through the center)	$2a$ -on y axis	$2a$ -on x axis
CONJUGATE AXIS (perpendicular to transverse axis at center)	$2b$, -on y axis	$2b$, -on x axis
CENTER	(h,k)	(h,k)

FOCI $(h \pm c, k)$ $(h, k \pm c)$

VERTICES $(h \pm a, k)$ $(h, k \pm a)$

Sample Problem:

Find all the identifying features of a hyperbola given its equation.

$$\frac{(x+3)^2}{4} - \frac{(y-4)^2}{16} = 1$$

Identify all known variables: $h = {}^-3 \quad k = 4 \quad a = 2 \quad b = 4$
$c = 2\sqrt{5}$

Slope of asymptotes: $\pm 4/2$ or ± 2
Transverse axis: 4 units long
Conjugate axis: 8 units long
Center: $({}^-3, 4)$
Foci: $({}^-3 \pm 2\sqrt{5}, 4)$
Vertices: $({}^-1, 4)$ and $({}^-5, 4)$

One way to **graph points** is in the **rectangular coordinate system**. In this system, the point (a, b) describes the point whose distance along the x-axis is "a" and whose distance along the y-axis is "b." The other method used to locate points is the **polar plane coordinate system**. This system consists of a fixed point called the pole or origin (labeled O) and a ray with O as the initial point called the polar axis. The ordered pair of a point P in the polar coordinate system is (r, θ), where $|r|$ is the distance from the pole and θ is the angle measure from the polar axis to the ray formed by the pole and point P. The coordinates of the pole are $(0, \theta)$, where θ is arbitrary. Angle θ can be measured in either degrees or in radians.

Sample problem:

1. Graph the point P with polar coordinates ($^-$2, $^-$45 degrees).

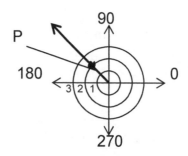

Draw $\theta = {}^-45$ degrees in standard position. Since r is negative, locate the point $\left|{}^-2\right|$ units from the pole on the ray opposite the terminal side of the angle. Note that P can be represented by ($^-$2, $^-$45 degrees + 180 degrees) = (2, 135 degrees) or by ($^-$2, $^-$45 degrees – 180 degrees) = (2, $^-$225 degrees).

2. Graph the point $P = \left(3, \dfrac{\pi}{4}\right)$ and show another graph that also represents the same point P.

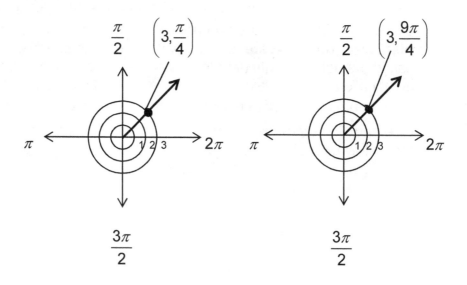

In the second graph, the angle 2π is added to $\dfrac{\pi}{4}$ to give the point $\left(3, \dfrac{9\pi}{4}\right)$.

It is possible that r be allowed to be negative. Now instead of measuring $|r|$ units along the terminal side of the angle, we would locate the point $|^-3|$ units from the pole on the ray opposite the terminal side. This would give the points $\left(^-3, \dfrac{5\pi}{4}\right)$ and $\left(^-3, \dfrac{^-3\pi}{4}\right)$.

We can represent any two-dimensional geometric figure in the **Cartesian** or **rectangular coordinate system**. The Cartesian or rectangular coordinate system is formed by two perpendicular axes (coordinate axes): the X-axis and the Y-axis. If we know the dimensions of a two-dimensional, or planar, figure, we can use this coordinate system to visualize the shape of the figure.

Example: Represent an isosceles triangle with two sides of length 4.

Draw the two sides along the x- and y- axes and connect the points (vertices).

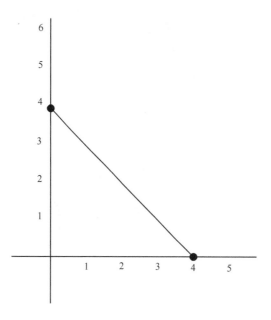

In order to represent three-dimensional figures, we need three coordinate axes (X, Y, and Z) which are all mutually perpendicular to each other. Since we cannot draw three mutually perpendicular axes on a two-dimensional surface, we use oblique representations.

Example: Represent a cube with sides of 2.

Once again, we draw three sides along the three axes to make things easier.

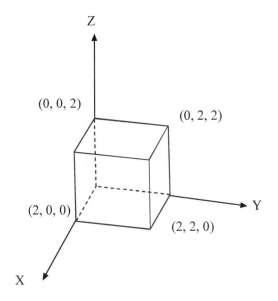

Each point has three coordinates (x, y, z).

The **equation of a circle** with its center at (h,k) and a radius r units is:

$$(x - h)^2 + (y - k)^2 = r^2$$

Sample Problem:

1. Given the equation $x^2 + y^2 = 9$, find the center and the radius of the circle. Then graph the equation.

First, writing the equation in standard circle form gives:

$$(x - 0)^2 + (y - 0)^2 = 3^2$$

therefore, the center is (0,0) and the radius is 3 units.

Sketch the circle:

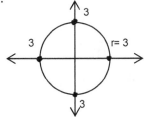

2. Given the equation $x^2 + y^2 - 3x + 8y - 20 = 0$, find the center and the radius. Then graph the circle.

First, write the equation in standard circle form by completing the square for both variables.

$x^2 + y^2 - 3x + 8y - 20 = 0$ 1. Complete the squares.

$(x^2 - 3x + 9/4) + (y^2 + 8y + 16) = 20 + 9/4 + 16$

$(x - 3/2)^2 + (y + 4)^2 = 153/4$

The center is $(3/2, {}^-4)$ and the radius is $\dfrac{\sqrt{153}}{2}$ or $\dfrac{3\sqrt{17}}{2}$.

Graph the circle.

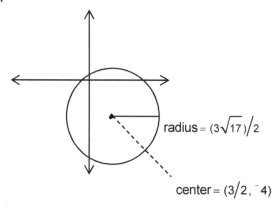

radius $= (3\sqrt{17})/2$

center $= (3/2, {}^-4)$

To **write the equation** given the center and the radius use the standard form of the equation of a circle:

$$(x - h)^2 + (y - k)^2 = r^2$$

Sample problems:

Given the center and radius, write the equation of the circle.

1. Center $({}^-1, 4)$; radius 11

$(x - h)^2 + (y - k)^2 = r^2$ 1. Write standard equation.

$(x - ({}^-1))^2 + (y - (4))^2 = 11^2$ 2. Substitute.

$(x + 1)^2 + (y - 4)^2 = 121$ 3. Simplify.

2. Center $(\sqrt{3}, {}^-1/2)$; radius $= 5\sqrt{2}$

$$(x - h)^2 + (y - k)^2 = r^2$$

1. Write standard equation.

$$(x - \sqrt{3})^2 + (y - ({}^-1/2))^2 = (5\sqrt{2})^2$$

2. Substitute.

$$(x - \sqrt{3})^2 + (y + 1/2)^2 = 50$$

3. Simplify.

0017. Apply mathematical principles and techniques to model and solve problems involving vector and transformational geometries.

Sample problem:

A pilot is traveling at an air speed of 300 mph and a direction of 20 degrees. Find the horizontal vector (ground speed) and the vertical vector (climbing speed).

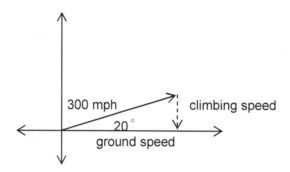

1. Draw sketch.
2. Use appropriate trigonometric ratio to calculate the component vectors.

To find the vertical vector: To find the horizontal vector:

$$\sin x = \frac{\text{opposite}}{\text{hypotenuse}}$$

$$\cos x = \frac{\text{adjacent}}{\text{hypotenuse}}$$

$$\sin(20) = \frac{c}{300}$$

$$\cos(20) = \frac{g}{300}$$

$$c = (.3420)(300)$$

$$g = (.9397)(300)$$

$$c = 102.606$$

$$g = 281.908$$

The **dot product** $a \cdot b$:

$$a = (a_1, a_2) = a_1 i + a_2 j \quad \text{and} \quad b = (b_1, b_2) = b_1 i + b_2 j$$

$$a \cdot b = a_1 b_1 + a_2 b_2$$

$a \cdot b$ is read "a dot b". Dot products are also called scalar or inner products. When discussing dot products, it is important to remember that "a dot b" is not a vector, but a real number.

Properties of the dot product:

$a \cdot a = |a|^2$

$a \cdot b = b \cdot a$

$a \cdot (b + c) = a \cdot b + a \cdot c$

$(ca) \cdot b = c(a \cdot b) = a \cdot (cb)$

$0 \cdot a = 0$

Sample problems:

Find the dot product.

1. $a = (5, 2), b = (^-3, 6)$

   $a \cdot b = (5)(^-3) + (2)(6)$

   $\quad = {}^-15 + 12$

   $\quad = {}^-3$

2. $a = (5i + 3j), b = (4i - 5j)$

   $a \cdot b = (5)(4) + (3)(^-5)$

   $\quad = 20 - 15$

   $\quad = 5$

3. The magnitude and direction of a constant force are given by $a = 4i + 5j$. Find the amount of work done if the point of application of the force moves from the origin to the point $P(7, 2)$.

The work W done by a constant force a as its point of application moves along a vector b is $W = a \cdot b$.

Sketch the constant force vector a and the vector b. Occasionally, it is important to reverse the addition or subtraction process and express the single vector as the sum or difference of two other vectors. It may be critically important for a pilot to understand not only the air velocity but also the ground speed and the climbing speed.

A **transformation** is a change in the position, shape, or size of a geometric figure. **Transformational geometry** is the study of manipulating objects by flipping, twisting, turning and scaling. **Symmetry** is exact similarity between two parts or halves, as if one were a mirror image of the other.

A **Tessellation** is an arrangement of closed shapes that completely covers the plane without overlapping or leaving gaps. Unlike **tilings**, tessellations do not require the use of regular polygons. In art the term is used to refer to pictures or tiles mostly in the form of animals and other life forms, which cover the surface of a plane in a symmetrical way without overlapping or leaving gaps. M. C. Escher is known as the "Father" of modern tessellations. Tessellations are used for tiling, mosaics, quilts and art.

If you look at a completed tessellation, you will see the original motif repeats in a pattern. There are 17 possible ways that a pattern can be used to tile a flat surface or "wallpaper."

There are four basic transformational symmetries that can be used in tessellations: **translation, rotation, reflection,** and **glide reflection**. The transformation of an object is called its image. If the original object was labeled with letters, such as $ABCD$, the image may be labeled with the same letters followed by a prime symbol, $A'B'C'D'$.

A **translation** is a transformation that "slides" an object a fixed distance in a given direction. The original object and its translation have the same shape and size, and they face in the same direction.

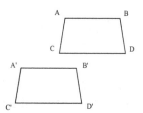

An example of a translation in architecture would be stadium seating. The seats are the same size and the same shape and face in the same direction.

A **rotation** is a transformation that turns a figure about a fixed point called the center of rotation. An object and its rotation are the same shape and size, but the figures may be turned in different directions. Rotations can occur in either a clockwise or a counterclockwise direction.

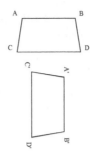

Rotations can be seen in wallpaper and art, and a Ferris wheel is an example of rotation.

An object and its **reflection** have the same shape and size, but the figures face in opposite directions.

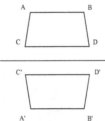

The line (where a mirror may be placed) is called the **line of reflection**. The distance from a point to the line of reflection is the same as the distance from the point's image to the line of reflection.

A **glide reflection** is a combination of a reflection and a translation.

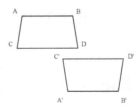

The tessellation below is a combination of the four types of transformational symmetry we have discussed:

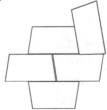

A **frieze** is a pattern that repeats in one direction. Friezes are often seen as ornaments in architecture. There are seven different frieze patterns possible: translation, glide reflection, two parallel reflections, two half turns, a reflection and a half turn, horizontal reflection, and three reflections.

Starting with this pattern

we derive these seven possibilities:

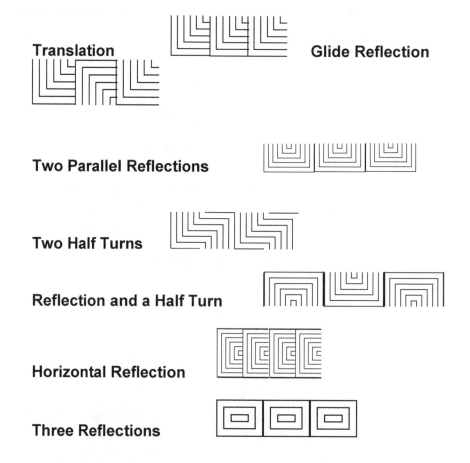

Translation **Glide Reflection**

Two Parallel Reflections

Two Half Turns

Reflection and a Half Turn

Horizontal Reflection

Three Reflections

A **fractal** is an endlessly repeating pattern that varies according to a set formula, a mixture of art and geometry. A fractal is any pattern that reveals greater complexity as it is enlarged. An example of a fractal is an ice crystal freezing on a glass window.

Fractals are **self-similar** and have **fractional (fractal) dimension**. Self-similar means that a fractal looks the same over all ranges of scale.

Example:

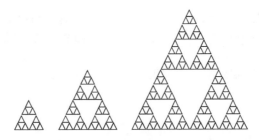

Fractional, or fractal dimension, means that the dimension of the figure is a non-integer, or fraction.

Example,

In the above figure, notice that the second triangle is composed of three miniature triangles exactly like the original. The length of any side of one of the miniature triangles could be multiplied by two to produce the entire triangle (S=2). The resulting figure consists of three separate identical miniature pieces (N=3). The formula to find the dimension of a strictly self-similar fractal is

$$D = \frac{\log N}{\log S}$$

For the above figure,

$$D = \frac{\log 3}{\log 2}$$

$$D = 1.585$$

Example:

Plot the given ordered pairs on a coordinate plane and join them in the given order, then join the first and last points.

(-3, -2), (3, -2), (5, -4), (5, -6), (2, -4), (-2, -4), (-5, -6), (-5, -4)

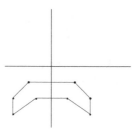

Increase all y-coordinates by 6.
(-3, 4), (3, 4), (5, 2), (5, 0), (2, 2), (-2, 2), (-5, 0), (-5, 2)

Plot the points and join them to form a second figure.

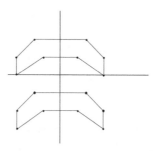

A figure on a coordinate plane can be translated by changing the ordered pairs.

A **transformation matrix** defines how to map points from one coordinate space into another coordinate space. The matrix used to accomplish two-dimensional transformations is described mathematically by a 3-by-3 matrix.

Example:
A point transformed by a 3-by-3 matrix

$$[x \quad y \quad 1] \times \begin{pmatrix} a & b & u \\ c & d & v \\ t_x & t_y & w \end{pmatrix} = [x' \quad y' \quad 1]$$

A 3-by-3 matrix transforms a point (x, y) into a point (x', y') by means of the following equations:

$$x' = ax + cy + t_x$$

$$y' = bx + dy + t_y$$

Another type of transformation is **dilation**. Dilation is a transformation that "shrinks" or "makes it bigger."

Example:

Using dilation to transform a diagram.

Starting with a triangle whose center of dilation is point P,

we dilate the lengths of the sides by the same factor to create a new triangle.

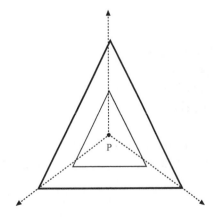

Example:

Using the vector method, prove the following: When two circles with centers X and Y intersect at points P and Q, XY is the perpendicular bisector of PQ.

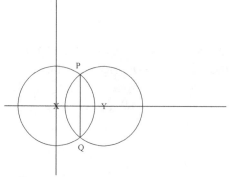

Let r represent a radius.

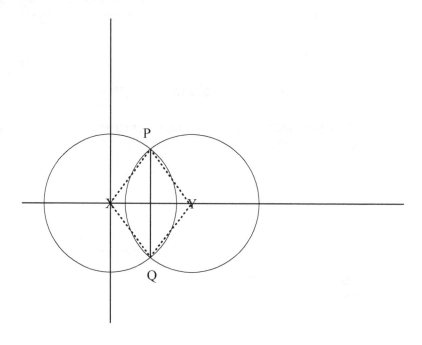

r1 = (x, y) = **XP**
r1* = (x, -y) = **XQ**
r2 = (r1+r2-x, -y) = **PY**
r2* = (r1+r2-x, y) = **QY**

XY is represented by vectors XP and PY = r1 + r2.
PQ is represented by vectors XQ and QY = r1 − r1*.

Using the vector dot product,

(**r1** + **r2**) dot (**r1** − **r1***) = (x + r1 + r2 − x, y − y) dot (x-x, y − (-y))
(**r1** + **r2**) dot (**r1** − **r1***) = (r1 + r2, 0) dot (0, 2y)
(**r1** + **r2**) dot (**r1** − **r1***) = 0

If the vector dot product is equal to zero, the two lines are perpendicular. Therefore, XY is perpendicular to PQ.
Example:

Using transformational methods, prove the Pythagorean Theorem.

Assume you have a right triangle with legs b and a, and hypotenuse c. Construct a segment subdivided into two parts of lengths a and b.

Using several rotations, construct a square on side a and a square on side b to create two regions whose total area is $a^2 + b^2$.

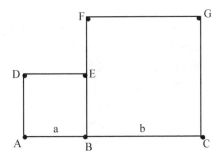

Define a translation from B to A and translate point C to get point H. Connect H to D and H to G, resulting in two right triangles.

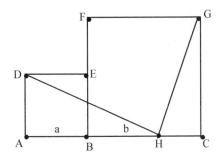

Hide segment BC and create segments BH and HC. This is so that we have well-defined triangle sides for the next step – rotating right triangle ADH 90 degrees about its top vertex, and right triangle HGC 90 degrees about its top vertex.

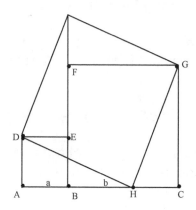

Note: Figure not drawn to scale.

Now prove that this construction yields a square (on DH) of side length c, and thus, since the area of this square is clearly equal to the sum of the areas of the original two squares, we have $a^2 + b^2 = c^2$, and the proof will be complete.

By SAS, triangle HCG must be congruent to the original right triangle, and thus its hypotenuse must be c. Also, by SAS, triangle DAH is also congruent to the original triangle, and so its hypotenuse is also c. Then, angles AHD and CHG (=ADH) must sum to 90 degrees, and the angle DHG is a right angle. Thus, you have shown that the construction yields a square on DH of side length c, and the proof is complete.

SUBAREA V–DATA ANALYSIS, PROBABILITY, STATISTICS, AND DISCRETE MATHEMATICS

0018. Understand the principles, properties, and techniques related to sequence, series, summation, and counting strategies and their applications to problem solving.

When given a set of numbers where the common difference between the terms is constant, use the following formula:

$$a_n = a_1 + (n-1)d$$

where a_1 = the first term

n = the n th term (general term)

d = the common difference

Sample problem:

1. Find the 8th term of the arithmetic sequence 5, 8, 11, 14, ...

$a_n = a_1 + (n-1)d$

$a_1 = 5$ Identify 1st term.

$d = 3$ Find d.

$a_8 = 5 + (8-1)3$ Substitute.

$a_8 = 26$

2. Given two terms of an arithmetic sequence find a and d.

$a_4 = 21$ $a_6 = 32$

$a_n = a_1 + (n-1)d$

$21 = a_1 + (4-1)d$

$32 = a_1 + (6-1)d$

$21 = a_1 + 3d$ Solve the system of equations.

$32 = a_1 + 5d$

$21 = a_1 + 3d$

$\dfrac{-32 = {}^-a_1 - 5d}{}$ Multiply by $^-1$ and add the

${}^-11 = {}^-2d$ equations.

$5.5 = d$

$21 = a_1 + 3(5.5)$ Substitute $d = 5.5$ into one of the

$21 = a_1 + 16.5$ equations.

$a_1 = 4.5$

The sequence begins with 4.5 and has a common difference of 5.5 between numbers.

When using geometric sequences consecutive numbers are compared to find the common ratio.

$$r = \frac{a_{n+1}}{a_n}$$

r = the common ratio
a_n = the nth term

The ratio is then used in the geometric sequence formula:

$$a_n = a_1 r^{n-1}$$

Sample problems:

1. Find the 8th term of the geometric sequence 2, 8, 32, 128 ...

$r = \dfrac{a_{n+1}}{a_n}$ Use the common ratio formula to find r.

$r = \dfrac{8}{2}$ Substitute $a_n = 2$ $a_{n+1} = 8$

$r = 4$

$a_n = a_1 \times r^{n-1}$ Use $r = 4$ to solve for the 8th term.
$a_8 = 2 \times 4^{8-1}$
$a_8 = 32768$

A basic Fibonacci sequence is when two numbers are added together to get the next number in the sequence. An example would be 1, 1, 2, 3, 5, 8, 13, ...

The difference between permutations and combinations is that in permutations all possible ways of writing an arrangement of objects are given while in a combination a given arrangement of objects is listed only once.

Given the set {1, 2, 3, 4}, list the arrangements of two numbers that can be written as a combination and as a permutation.

Combination	Permutation
12, 13, 14, 23, 24, 34	12, 21, 13, 31, 14, 41, 23, 32, 24, 42, 34, 43,
six ways	twelve ways

Using the formulas given below the same results can be found.

$$_nP_r = \frac{n!}{(n-r)!}$$

The notation $_nP_r$ is read "the number of permutations of n objects taken r at a time."

$$_4P_2 = \frac{4!}{(4-2)!}$$

Substitute known values.

$$_4P_2 = 12$$

Solve.

$$_nC_r = \frac{n!}{(n-r)!r!}$$

The number of combinations when r objects are selected from n objects.

$$_4C_2 = \frac{4!}{(4-2)!2!}$$

Substitute known values.

$$_4C_2 = 6$$

Solve.

The binomial expansion theorem is another method used to find the coefficients of $(x+y)$. Although Pascal's Triangle is easy to use for small values of n, it can become cumbersome to use with larger values of n.

Binomial Theorem:

For any positive value of n,

$$(x+y)^n = x^n + \frac{n!}{(n-1)!1!}x^{n-1}y + \frac{n!}{(n-2)!2!}x^{n-2}y^2 + \frac{n!}{(n-3)!3!}x^{n-3}y^3 + \frac{n!}{1!(n-1)!}xy^{n-1} + y^n$$

Sample Problem:

1. Expand $(3x+y)^5$

$$(3x)^5 + \frac{5!}{4!1!}(3x)^4 y^1 + \frac{5!}{3!2!}(3x)^3 y^2 + \frac{5!}{2!3!}(3x)^2 y^3 + \frac{5!}{1!4!}(3x)^1 y^4 + y^5 =$$

$$243x^5 + 405x^4y + 270x^3y^2 + 90x^2y^3 + 15xy^4 + y^5$$

Any term of a binomial expansion can be written individually. For example, the y value of the seventh term of $(x+y)^n$, would be raised to the 6th power and since the sum of exponents on x and y must equal seven, then the x must be raised to the $n-6$ power.

The formula to find the rth term of a binomial expansion is:

$$\frac{n!}{[n-(r-1)]!(r-1)!}x^{n-(r-1)}y^{r-1}$$

where $r=$ the number of the desired term and $n=$ the power of the binomial.

Sample Problem:

1. Find the third term of $(x+2y)^{11}$

$x^{n-(r-1)}$ $\qquad$ y^{r-1} $\qquad\qquad$ Find x and y exponents.

$x^{11-(3-1)}$ $\qquad$ y^{3-1}

x^9 $\qquad\qquad$ y^2

$\frac{11!}{9!2!}(x^9)(2y)^2$ $\qquad\qquad$ Substitute known values.

$220x^9y^2$ $\qquad\qquad$ Solve.

Practice problems:

1. $(x+y)^7$; 5th term

2. $(3x-y)^9$; 3rd term

0019. Understand the principles, properties, and techniques of probability and their applications.

Percentiles divide data into 100 equal parts. A person whose score falls in the 65th percentile has outperformed 65 percent of all those who took the test. This does not mean that the score was 65 percent out of 100 nor does it mean that 65 percent of the questions answered were correct. It means that the grade was higher than 65 percent of all those who took the test.

Stanine "standard nine" scores combine the understandability of percentages with the properties of the normal curve of probability. Stanines divide the bell curve into nine sections, the largest of which stretches from the 40th to the 60th percentile and is the "Fifth Stanine" (the average of taking into account error possibilities).

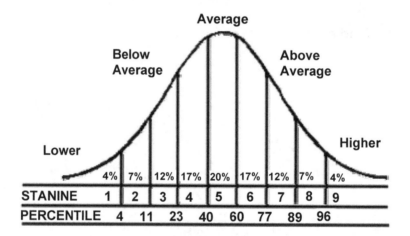

Quartiles divide the data into 4 parts. First find the median of the data set (Q2), then find the median of the upper (Q3) and lower (Q1) halves of the data set. If there are an odd number of values in the data set, include the median value in both halves when finding quartile values. For example, given the data set: {1, 4, 9, 16, 25, 36, 49, 64, 81} first find the median value, which is 25 this is the second quartile. Since there are an odd number of values in the data set (9), we include the median in both halves. To find the quartile values, we much find the medians of: {1, 4, 9, 16, 25} and {25, 36, 49, 64, 81}. Since each of these subsets had an odd number of elements (5), we use the middle value. Thus the first quartile value is 9 and the third quartile value is 49. If the data set had an even number of elements, average the middle two values. The quartile values are always either one of the data points, or exactly half way between two data points.

Sample problem:

1. Given the following set of data, find the percentile of the score 104.

> 70, 72, 82, 83, 84, 87, 100, 104, 108, 109, 110, 115

Solution: Find the percentage of scores below 104.

7/12 of the scores are less than 104. This is 58.333%; therefore, the score of 104 is in the 58th percentile.

2. Find the first, second and third quartile for the data listed.
6, 7, 8, 9, 10, 12, 13, 14, 15, 16, 18, 23, 24, 25, 27, 29, 30, 33, 34, 37

Quartile 1: The 1st Quartile is the median of the lower half of the data set, which is 11.

Quartile 2: The median of the data set is the 2nd Quartile, which is 17.

Quartile 3: The 3rd Quartile is the median of the upper half of the data set, which is 28.
Different situations require different information. If we examine the circumstances under which an ice cream store owner may use statistics collected in the store, we find different uses for different information.

Over a 7-day period, the store owner collected data on the ice cream flavors sold. He found the mean number of scoops sold was 174 per day. The most frequently sold flavor was vanilla. This information was useful in determining how much ice cream to order in all and in what amounts for each flavor.

In the case of the ice cream store, the median and range had little business value for the owner.

Consider the set of test scores from a math class: 0, 16, 19, 65, 65, 65, 68, 69, 70, 72, 73, 73, 75, 78, 80, 85, 88, and 92. The mean is 64.06 and the median is 71. Since there are only three scores less than the mean out of the eighteen scores, the median (71) would be a more descriptive score.

Retail store owners may be most concerned with the most common dress size so they may order more of that size than any other.

An understanding of the definitions is important in determining the validity and uses of statistical data. All definitions and applications in this section apply to ungrouped data.

Data item: each piece of data is represented by the letter X.

Mean: the average of all data represented by the symbol $\overline{X}$.

Range: difference between the highest and lowest value of data items.

Sum of the Squares: sum of the squares of the differences between each item and the mean.

$$Sx^2 = (X - \overline{X})^2$$

Variance: the sum of the squares quantity divided by the number of items. (the lower case Greek letter sigma squared (σ^2)represents variance).

$$\frac{Sx^2}{N} = \sigma^2$$

The larger the value of the variance the larger the spread

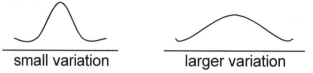

small variation   larger variation

Standard Deviation: the square root of the variance. The lower case Greek letter sigma (σ) is used to represent standard deviation. $\sigma = \sqrt{\sigma^2}$

Most statistical calculators have standard deviation keys on them and should be used when asked to calculate statistical functions. It is important to become familiar with the calculator and the location of the keys needed.

Basic statistical concepts can be applied without computations. For example, inferences can be drawn from a graph or statistical data. A bar graph could display which grade level collected the most money. Student test scores would enable the teacher to determine which units need to be remediated.

The Addition Principle of Counting states:

If A and B are events, $n(AorB) = n(A) + n(B) - n(A \cap B)$.

Example:

In how many ways can you select a black card or a Jack from an ordinary deck of playing cards?

Let B denote the set of black cards and let J denote the set of Jacks. Then, $n(B) = 26, n(J) = 4, n(B \cap J) = 2$ and

$$n(BorJ) = n(B) + n(J) - n(B \cap A)$$
$$= 26 + 4 - 2$$
$$= 28.$$

The Addition Principle of Counting for Mutually Exclusive Events states:

If A and B are mutually exclusive events, $n(AorB) = n(A) + n(B)$.

Example:

A travel agency offers 40 possible trips: 14 to Asia, 16 to Europe and 10 to South America. In how many ways can you select a trip to Asia or Europe through this agency?

Let A denote trips to Asia and let E denote trips to Europe. Then, $A \cap E = \varnothing$ and

$$n(AorE) = 14 + 16 = 30.$$

Therefore, the number of ways you can select a trip to Asia or Europe is 30.

The Multiplication Principle of Counting for Dependent Events states:
Let A be a set of outcomes of Stage 1 and B a set of outcomes of Stage 2.

Then the number of ways $n(AandB)$, that A and B can occur in a two-stage experiment is given by:

$$n(AandB) = n(A)n(B|A),$$

where $n(B|A)$ denotes the number of ways B can occur given that A has already occurred.

Example:

How many ways from an ordinary deck of 52 cards can two Jacks be drawn in succession if the first card is drawn but not replaced in the deck and then the second card is drawn?

This is a two-stage experiment for which we wish to compute $n(AandB)$, where A is the set of outcomes for which a Jack is obtained on the first draw and B is the set of outcomes for which a Jack is obtained on the second draw.

If the first card drawn is a Jack, then there are only three remaining Jacks left to choose from on the second draw. Thus, drawing two cards without replacement means the events A and B are dependent.

$$n(AandB) = n(A)n(B|A) = 4 \cdot 3 = 12$$

The Multiplication Principle of Counting for Independent Events states:
Let A be a set of outcomes of Stage 1 and B a set of outcomes of Stage 2. If A and B are independent events then the number of ways $n(AandB)$, that A and B can occur in a two-stage experiment is given by:

$$n(AandB) = n(A)n(B).$$

Example:

How many six-letter code "words" can be formed if repetition of letters is not allowed?

Since these are code words, a word does not have to look like a word; for example, abcdef could be a code word. Since we must choose a first letter *and* a second letter *and* a third letter *and* a fourth letter *and* a fifth letter *and* a sixth letter, this experiment has six stages.

Since repetition is not allowed there are 26 choices for the first letter; 25 for the second; 24 for the third; 23 for the fourth; 22 for the fifth; and 21 for the sixth. Therefore, we have:

n(six-letter code words without repetition of letters)

$$= 26 \cdot 25 \cdot 24 \cdot 23 \cdot 22 \cdot 21$$

$$= 165,765,600$$

A **Bernoulli trial** is an experiment whose outcome is random and can be either of two possible outcomes, called "success" or "failure." Tossing a coin would be an example of a Bernoulli trial. We make the outcomes into a random variable by assigning the number 0 to one outcome and the number 1 to the other outcome. Traditionally, the "1" outcome is considered the "success" and the "0" outcome is considered the "failure." The probability of success is represented by p, with the probability of failure being $1-p$, or q.

Bernoulli trials can be applied to any real-life situation in which there are just two possible outcomes. For example, concerning the birth of a child, the only two possible outcomes for the sex of the child are male or female.

0020. Understand the principles, properties, and techniques of data analysis and statistics.

Dependent events occur when the probability of the second event depends on the outcome of the first event. For example, consider the two events (A) it is sunny on Saturday and (B) you go to the beach. If you intend to go to the beach on Saturday, rain or shine, then A and B may be independent. If however, you plan to go to the beach only if it is sunny, then A and B may be dependent. In this situation, the probability of event B will change depending on the outcome of event A.

Suppose you have a pair of dice, one red and one green. If you roll a three on the red die and then roll a four on the green die, we can see that these events do not depend on the other. The total probability of the two independent events can be found by multiplying the separate probabilities.

$$P(A \text{ and } B) = P(A) \times P(B)$$
$$= 1/6 \times 1/6$$
$$= 1/36$$

Many times, however, events are not independent. Suppose a jar contains 12 red marbles and 8 blue marbles. If you randomly pick a red marble, replace it and then randomly pick again, the probability of picking a red marble the second time remains the same. However, if you pick a red marble, and then pick again without replacing the first red marble, the second pick becomes dependent upon the first pick.

$$P(\text{Red and Red}) \text{ with replacement} = P(\text{Red}) \times P(\text{Red})$$
$$= 12/20 \times 12/20$$
$$= 9/25$$

$$P(\text{Red and Red}) \text{ without replacement} = P(\text{Red}) \times P(\text{Red})$$
$$= 12/20 \times 11/19$$
$$= 33/95$$

Mean, median and mode are three measures of central tendency. The **mean** is the average of the data items. The **median** is found by putting the data items in order from smallest to largest and selecting the item in the middle (or the average of the two items in the middle). The **mode** is the most frequently occurring item.

Range is a measure of variability. It is found by subtracting the smallest value from the largest value.

Sample problem:

Find the mean, median, mode and range of the test score listed below:

85	77	65
92	90	54
88	85	70
75	80	69
85	88	60
72	74	95

Mean (X) = sum of all scores ÷ number of scores
         = 78

Median = put numbers in order from smallest to largest. Pick middle number.
54, 60, 65, 69, 70, 72, 74, 75, 77, 80, 85, 85, 85, 88, 88, 90, 92, 95
                              -- --
                         both in middle

Therefore, median is average of two numbers in the middle or 78.5

Mode = most frequent number
     = 85

Range = largest number minus the smallest number
      = 95 − 54
      = 41

Different situations require different information. If we examine the circumstances under which an ice cream store owner may use statistics collected in the store, we find different uses for different information.

Over a 7-day period, the store owner collected data on the ice cream flavors sold. He found the mean number of scoops sold was 174 per day. The most frequently sold flavor was vanilla. This information was useful in determining how much ice cream to order in all and in what amounts for each flavor.

In the case of the ice cream store, the median and range had little business value for the owner.

Consider the set of test scores from a math class: 0, 16, 19, 65, 65, 65, 68, 69, 70, 72, 73, 73, 75, 78, 80, 85, 88, and 92. The mean is 64.06 and the median is 71. Since there are only three scores less than the mean out of the eighteen score, the median (71) would be a more descriptive score.

Retail store owners may be most concerned with the most common dress size so they may order more of that size than any other.

The **t-test** is the most commonly used method to evaluate the difference in means between two groups. The t-test assesses whether the means of two groups are statistically different from each other. The formula for the t-test is a ratio. The top part of the ratio is the difference between the two means or averages. The bottom part is a measure of the variability or dispersion of the scores.

$$\frac{\text{difference between group means}}{\text{variability of groups}} = \frac{\overline{X}_T - \overline{X}_C}{SE(\overline{X}_T - \overline{X}_C)} = \text{t-value}$$

In this example T refers to a treatment group and C refers to a control group.

To compute the top part of the formula, just find the difference between the means. The bottom part of the formula is called the **standard error of the difference**. To compute it, take the variance for each group and divide it by the number of people in that group. Add these two values and then take their square root.

$$SE(\overline{X}_T - \overline{X}_C) = \sqrt{\frac{\text{var}_T}{n_T} + \frac{\text{var}_C}{n_C}}$$

The final formula for the t-test is

$$t = \frac{\overline{X}_T - \overline{X}_C}{\sqrt{\frac{\text{var}_T}{n_T} + \frac{\text{var}_C}{n_C}}}$$

Once you compute the t-value, you have to look it up in a table of significance to test whether the ratio is large enough to say that the difference between the groups is not likely to have been a chance finding. To test the significance, you need to set a risk, or alpha level, which is usually .05. You also must determine the degrees of freedom (df) for the test. In the t-test, the df is the sum of the persons in both groups minus 2. Using the t-value, the alpha level, and the df, you can look up the t-value in a standard table of significance to determine whether the t-value is large enough to be significant. If it is, you can conclude that the difference between the means for the two groups is different.

The **chi-square test** is a method of determining the odds for or against a given deviation from expected statistical distribution.

Example:

We want to determine if the odds of flipping a coin heads-up is the same as tails-up; is the coin flipped fairly? We collect data by flipping the coin 200 times. The coin landed heads-up 92 times and tails-up 108 times.

To perform a chi-square test we first must establish a null hypothesis. In this example, the null hypothesis states that the coin should be equally likely to land head-up or tails-up, every time. The null hypothesis allows us to state expected frequencies. For 200 tosses we would expect 100 heads and 100 tails.

Next, prepare a table:

	Heads	Tails	Total
Observed	92	108	200
Expected	100	100	200
Total	192	208	400

The observed values are the data gathered. The expected values are the frequencies expected, based on the null hypothesis.

We calculate chi-squared:

$$\text{Chi-squared} = \frac{(\text{observed-expected})^2}{(\text{expected})}$$

We have two classes to consider in this example, heads and tails.

$$\text{Chi-squared} = \frac{(100-92)^2}{100} + \frac{(100-108)^2}{100}$$

$$= \frac{(8)^2}{100} + \frac{(-8)^2}{100}$$

$$= 0.64 + 0.64$$

$$= 1.28$$

We then consult a table of critical values of the chi-squared distribution. Here is a portion of such a table:

df/prob.	0.99	0.95	0.90	0.80	0.70	0.50	0.30	0.20	0.10	0.05
1	0.00013	0.0039	0.016	0.64	0.15	0.46	1.07	1.64	2.71	3.84
2	0.02	0.10	0.21	0.45	0.71	1.39	2.41	3.22	4.60	5.99
3	0.12	0.35	0.58	1.99	1.42	2.37	3.66	4.64	6.25	7.82
4	0.3	0.71	1.06	1.65	2.20	3.36	4.88	5.99	7.78	9.49
5	0.55	1.14	1.61	2.34	3.00	4.35	6.06	7.29	9.24	11.07

We determine the degrees of freedom (df) by subtracting one from the number of classes. In this example we have two classes (heads and tails), so df is 1. Our chi-squared value is 1.28. In the table our value lies between 1.07 (a probability of .30) and 1.64 (a probability of .20). Interpolation gives us an estimated probability of 0.26. This value means that there is a 74% chance that the coin is biased. Because the chi-squared value we obtained is greater than 0.05 (0.26 to be exact), we accept the null hypothesis as true and conclude that the coin is fair.

Correlation is a measure of association between two variables. It varies from -1 to 1, with 0 being a random relationship, 1 being a perfect positive linear relationship, and -1 being a perfect negative linear relationship.

The **correlation coefficient** (r) is used to describe the strength of the association between the variables and the direction of the association.

Example:

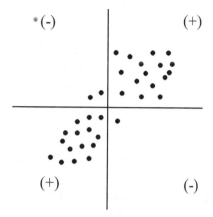

Horizontal and vertical lines are drawn through the point of averages which is the point on the averages of the x and y values. This divides the scatter plot into four quadrants. If a point is in the lower left quadrant, the product of two negatives is positive; in the upper right, the product of two positives is positive. The positive quadrants are depicted with the positive sign (+). In the two remaining quadrants (upper left and lower right), the product of a negative and a positive is negative. The negative quadrants are depicted with the negative sign (-). If r is positive, then there are more points in the positive quadrants and if r is negative, then there are more points in the two negative quadrants.

Regression is a form of statistical analysis used to predict a dependent variable (y) from values of an independent variable (x). A regression equation is derived from a known set of data.

The simplest regression analysis models the relationship between two variables using the following equation: $y = a + bx$, where y is the dependent variable and x is the independent variable. This simple equation denotes a linear relationship between x and y. This form would be appropriate if, when you plotted a graph of x and y, you tended to see the points roughly form along a straight line.

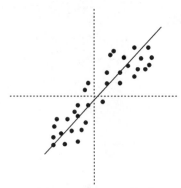

The line can then be used to make predictions.

If all of the data points fell on the line, there would be a perfect correlation ($r = 1.0$) between the x and y data points. These cases represent the best scenarios for prediction. A positive or negative r value represents how y varies with x. When r is positive, y increases as x increases. When r is negative y decreases as x increases.

A **linear regression** equation is of the form: $Y = a + bX$.

Example:

A teacher wanted to determine how a practice test influenced a student's performance on the actual test. The practice test grade and the subsequent actual test grade for each student are given in the table below:

Practice Test (x)	Actual Test (y)
94	98
95	94
92	95
87	89
82	85
80	78
75	73
65	67
50	45
20	40

We determine the equation for the linear regression line to be
$y = 14.650 + 0.834x$.

A new student comes into the class and scores 78 on the practice test. Based on the equation obtained above, what would the teacher predict this student would get on the actual test?

$$y = 14.650 + 0.834(78)$$
$$y = 14.650 + 65.052$$
$$y = 80$$

A **logarithmic regression** equation is of the form: $y = a + b \ln x$.

Example:

The water, w, in an open container is evaporating. The number of ounces remaining after h hours is shown in the table below:

Hours (h)	2	5	10	15	19	30
Water (w)	13	11	9	8.5	7.5	6.5

We construct a scatter plot for this data and find a logarithmic regression equation to model the data.

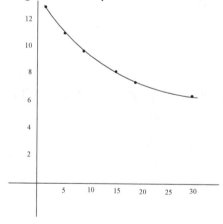

$$w = 14.71 - 2.41 \ln h$$

Using this regression equation predict how many ounces of water are remaining in the container after 48 hours.

$$w = 14.71 - 2.41\ln(48)$$
$$w = 14.71 - 2.41(3.87)$$
$$w = 14.71 - 9.33$$
$$w = 5.4$$

An **exponential regression** equation is of the form: $y = a(b)^x$. An exponential function may be used to model either growth or decay, depending on the value of b. When $b > 1$ the function models exponential growth. When $0 < b < 1$ the function models exponential decay. You may fit an exponential curve to data and find the exponential function.

Example:

For the data in the table below, determine the exponential regression equation.

x	y
-5	10.22
1	7.98
2	6.63
5	5.51
8	4.57

$$y = 7.502(.94)^x$$

Using this equation predict what y will be when $x = 12$.
$$y = 7.502(.94)^{12}$$
$$y = 7.502(.476)$$
$$y = 3.57$$

Power regression is a functional form used for nonlinear regression. The formula is: $Y = ab^X$.

Example:

On a newly discovered planet the weight, w, of an object and the distance, d, this object is from the surface of the planet were recorded and are shown in the table below (Weight is in pounds and distance in miles).

d	w
2	116
3	58
4	36
5	24
6	18

We construct a scatter plot for the given data using distance as the independent variable (x) and find a power regression equation to model this data.

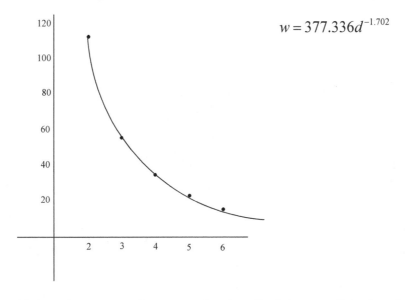

$$w = 377.336d^{-1.702}$$

Using the regression equation we find the predicted weight of the object to the nearest pound when it is 20 miles from the surface of the planet.

$$w = 377.336(20)^{-1.702}$$
$$= 377.336(.006)$$
$$= 2$$

Example – Life Science

Examine an animal population and vegetation density in a biome over time.

Example – Physical Science

Explore motions and forces by calculating speeds based on distance and time traveled and creating a graph to represent the data.

Example – Geography

Explore and illustrate knowledge of earth landforms.

Example – Economics/Finance

Compare car buying with car leasing by graphing comparisons and setting up monthly payment schedules based on available interest rates.

0021. Understand how techniques of discrete mathematics (e.g., diagrams, graphs, matrices, propositional statements) are applied in the analysis, and interpretation, communication, and solution of problems.

There are many graphical ways in which to represent data: line plots, line graphs, scatter plots, stem and leaf plots, histograms, bar graphs, pie charts, and pictographs.

A **line plot** organizes data in numerical order along a number line. An x is placed above the number line for each occurrence of the corresponding number. Line plots allow you to see at a glance a range of data and where typical and atypical data falls. They are generally used to summarize relatively small sets of data.

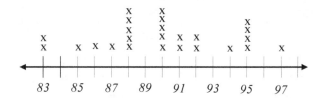

A **line graph** compares two variables, and each variable is plotted along an axis. A line graph highlights trends by drawing connecting lines between data points. They are particularly appropriate for representing data that varies continuously. Line graphs are sometimes referred to as **frequency polygons**.

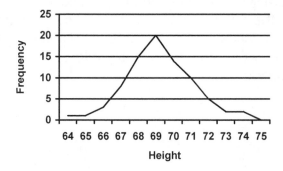

Scatter plots compare two characteristics of the same group of things or people and usually consist of a large body of data. They show how much one variable is affected by another. The relationship between the two variables is their **correlation**. The closer the data points come to making a straight line when plotted, the closer the correlation.

Stem and leaf plots are visually similar to line plots. The **stems** are the digits in the greatest place value of the data values, and the **leaves** are the digits in the next greatest place values. Stem and leaf plots are best suited for small sets of data and are especially useful for comparing two sets of data. The following is an example using test scores:

4	9
5	4 9
6	1 2 3 4 6 7 8 8
7	0 3 4 6 6 6 7 7 7 8 8 8 8
8	3 5 5 7 8
9	0 0 3 4 5
10	0 0

Histograms are used to summarize information from large sets of data that can be naturally grouped into intervals. The vertical axis indicates **frequency** (the number of times any particular data value occurs), and the horizontal axis indicates data values or ranges of data values. The number of data values in any interval is the **frequency of the interval**.

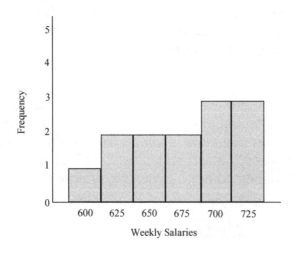

A **pictograph** uses small figures or icons to represent data. Pictographs are used to summarize relative amounts, trends, and data sets. They are useful in comparing quantities.

Monarch Butterfly Migration to the U.S.
in millions

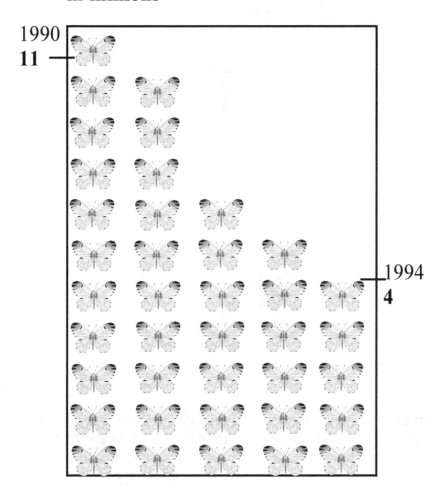

The data in this graph is not accurate. It is for illustration purposes only.

Bar graphs are similar to histograms. However, bar graphs are often used to convey information about categorical data where the horizontal scale represents a non-numeric attributes such as cities or years. Another difference is that the bars in bar graphs rarely touch. Bar graphs are also useful in comparing data about two or more similar groups of items.

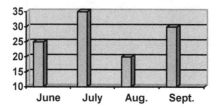

A **pie chart**, also known as a **circle graph**, is used to represent relative amounts of a whole.

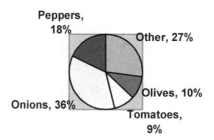

An **arithmetic sequence** is a sequence where each successive term is obtained from the previous term by addition or subtraction of a fixed number called a **difference**. In order for it to be an arithmetic sequence the SAME number must be added, or subtracted, if that is the pattern, each time. An example would be 1, 5, 9, 13, ... where 4 is added to each previous number.

A **geometric sequence** is a sequence where each successive term is obtained from the previous term by multiplying by a fixed number called a **ratio**. In order for it to be a geometric sequence, each term must be multiplied by the SAME number. An example would be 2, 8, 32, 128, ... where each term is multiplied by 4.

Sequences can be **finite** or **infinite**. A finite sequence is a sequence whose domain consists of the set {1, 2, 3, ... *n*} or the first *n* positive integers. An infinite sequence is a sequence whose domain consists of the set {1, 2, 3, ...}; which is in other words all positive integers.

A **recurrence relation** is an equation that defines a sequence recursively; in other words, each term of the sequence is defined as a function of the preceding terms.

A real-life application would be using a recurrence relation to determine how much your savings would be in an account at the end of a certain period of time. For example:

You deposit $5,000 in your savings account. Your bank pays 5% interest compounded annually. How much will your account be worth at the end of 10 years?

Let *V* represent the amount of money in the account and V_n represent the amount of money after *n* years.

The amount in the account after *n* years equals the amount in the account after *n* – 1 years plus the interest for the *n*th year. This can be expressed as the recurrence relation V_0 where your initial deposit is represented by $V_0 = 5,000$.

$$V_0 = V_0$$
$$V_1 = 1.05V_0$$
$$V_2 = 1.05V_1 = (1.05)^2 V_0$$
$$V_3 = 1.05V_2 = (1.05)^3 V_0$$
$$......$$
$$V_n = (1.05)V_{n-1} = (1.05)^n V_0$$

Inserting the values into the equation, you get
$V_{10} = (1.05)^{10}(5,000) = 8,144$.

You determine that after investing $5,000 in an account earning 5% interest, compounded annually for 10 years, you would have $8,144.

Graphs display data so that the data can be interpreted. Graphs are often used to see trends and predict future performance.

For example, this line graph depicts the auto sales for a car dealership. The car dealership is able to see at a glance how many cars were sold in a particular month and which months tended to have the least and greatest sales. This information helps him to control his inventory, forecast his sales, and manage his staffing. He might also use the information to plan ways in which to boost sales in lagging months.

AUTO SALES

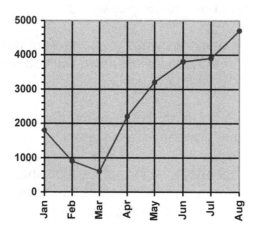

A **matrix** is an ordered set of numbers in rectangular form.

$$\begin{pmatrix} 0 & 3 & 1 \\ 4 & 2 & 3 \\ 1 & 0 & 2 \end{pmatrix}$$

Since this matrix has 3 rows and 3 columns, it is called a 3 x 3 matrix. The element in the second row, third column would be denoted as $3_{2,3}$.

Matrices are used often to solve systems of equations. They are also used by physicists, mathematicians, and biologists to organize and study data such as population growth. It is also used in finance for such purposes as investment growth analysis and portfolio analysis.

Matrices are easily translated into computer code in high-level programming languages and can be easily expressed in electronic spreadsheets.

A simple financial example of using a matrix to solve a problem follows:

A company has two stores. The income and expenses (in dollars) for the two stores, for three months, are shown in the matrices.

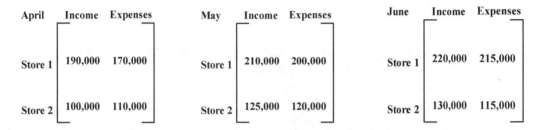

The owner wants to know what his first-quarter income and expenses were, so he adds the three matrices.

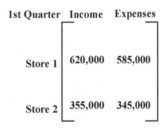

Then, to find the profit for each store:

Profit for Store 1 = $620,000 - $585,000 = $35,000
Profit for Store 2 = $355,000 - $345,000 = $10,000

An **algorithm** is a method of calculating; simply put, it can be multiplication, subtraction, or a combination of operations. When working with computers and calculators we employ **algorithmic thinking**, which means performing mathematical tasks by creating a sequential and often repetitive set of steps. A simple example would be to create an algorithm to generate the Fibonacci numbers utilizing the MR and M+ keys found on most calculators. The table below shows Entry made in the calculator, the value x seen in the display, and the value M contained in the memory.

Entry	ON/AC	1	M+	+	M+	MR	+	M+	MR	+	...
x	0	1	1	1	1	2	3	3	5	8	...
M	0	0	1	1	2	2	2	5	5	5	...

This eliminates the need to repeatedly enter required numbers.

Computers have to be programmed and many advanced calculators are programmable. A **program** is the steps of an algorithm that are entered into a computer or calculator. The main advantage of using a program is that once the algorithm is entered, a result may be obtained by merely hitting a single keystroke to select the program, thereby eliminating the need to continually enter a large number of steps. Teachers find that programmable calculators are excellent for investigating "what if?" situations.

Using graphing calculators or computer software has many advantages. The technology is better able to handle large data sets, such as the results of a science experiment and it is much easier to edit and sort the data and change the style of the graph to find its best representation. Furthermore, graphing calculators also provide a tool to plot statistics.

SUBAREA VI –ALGEBRA: CONSTRUCTED-RESPONSE ASSIGNMENT

Please see page 311 in the sample test solution section for an example of the constructed response assignment.

CURRICULUM AND INSTRUCTION

TEACHING METHODS - The art and science specific for high school mathematics

Some commonly used teaching techniques and tools are described below along with links to further information. The links provided provide a wealth of instructional ideas and materials. You should consider joining The National Council of Teachers of Mathematics as they have many ideas in their journals about pedagogy and curriculum standards and publish professional books that are useful. You can write to them at 1906 Association Drive, Reston, VA 20191-1593. You can also order a starter kit from them for $9 that includes 3 recent journals by calling 800-235-7566 or writing e-mailorders@nctm.org.

> A couple of resources for students to use at home: http://www.algebra.com/, http://www.mathsisfun.com/algebra/index.html and http://www.purplemath.com/. A helpful .pdf guide for parents is available at: http://my.nctm.org/ebusiness/ProductCatalog/product.aspx?ID=12931 A good website for understanding the causes of and how to prevent "math anxiety:" http://www.mathgoodies.com/articles/math_anxiety.html

1. Classroom warm-up: Engage your students as soon as they walk in the door: provide an interesting short activity each day. You can make use of thought-provoking questions and puzzles. Also use relevant puzzles specific to topics you may be covering in your class. The following websites provide some ideas:

    http://www.math-drills.com/?gclid=CP-P0dzenJICFRSTGgodNjG0Zw
    http://www.mathgoodies.com/games/
    http://mathforum.org/k12/k12puzzles/
    http://mathforum.org/pow/other.html

2. Real life examples: Connect math to other aspects of your students' lives by using examples and data from the real world whenever possible. It will not only keep them engaged, it will also help answer the perennial question "Why do we have to learn math?" Online resources to get you started:

    http://chance.dartmouth.edu/chancewiki/index.php/Main_Page has some interesting real-world probability problems (such as, "Can statistics determine if Robert Clemens used steroids?")
    http://www.mathnotes.com/nos_index.html has all kinds of links between math and the real world suitable for high school students
    http://www.nssl.noaa.gov/edu/ideas/ uses weather to teach math

http://standards.nctm.org/document/eexamples/index.htm#9-12 Using Graphs, Equations, and Tables to Investigate the Elimination of Medicine from the Body: Modeling the Situation
http://mathforum.org/t2t/faq/election.html Election math in the classroom
http://www.education-world.com/a_curr/curr148.shtml offers examples of real-life problems such as calculating car payments, saving and investing, the world of credit cards, and other finance problems.
http://score.kings.k12.ca.us/real.world.html is a website connecting math to real jobs, elections, NASA projects, etc.

3. Graphing and spreadsheets for enhancing math learning:
   http://www.cvgs.k12.va.us/digstats/
   http://score.kings.k12.ca.us/standards/probability.html for graphing and statistics.
   http://www.microsoft.com/education/solving.mspx for using spreadsheets and to solve polynomial problems.

4. Use technology - manipulatives, software and interactive online activities that can help all students learn, particularly those oriented more towards visual and kinesthetic learning. Here are some websites:
   http://illuminations.nctm.org/ActivitySearch.aspx has games for grades 9-12 that can be played against the computer or another student.
   http://nlvm.usu.edu/ The National Library of Virtual Manipulatives has resources for all grades on numbers and operations, algebra, geometry, probability and measurement.
   http://mathforum.org/pow/other.html has links to various math challenges, manipulatives and puzzles.
   http://www.etacuisenaire.com/algeblocks/algeblocks.jsp Algeblocks are blocks that utilize the relationship between algebra and geometry.

5. Word problem strategies- the hardest thing to do is take the English and turn it into math but there are 6 key steps to teach students how to solve word problems:

   a. The problem will have **key words** to suggest the type of operation or operations to be performed to solve the problem. For example, words such as "altogether" or "total" imply addition while words such as "difference" or "How many more?" imply subtraction.
   b. **Pictures or Concrete Materials**: Math is very abstract; it is easier to solve a problem using pictures or concrete materials to illustrate the problem. Pictures and concrete materials allow the students to manipulate the material to solve the problem with trial and error. Model drawing pictures and using concrete materials to solve word problems.

c. *Use Logic*: Ask your students if their answers make sense. Get them used to using the process of deduction. Model the deduction process for them to decide on the answer to a word problem. For example, in solving a problem such as: Two consecutive numbers have a sum of 91. What are the numbers? If the student arrives at an answer of 44 and 45 it is obvious that there was an error in the equation used or calculation since 44 and 45 are consecutive but don't add up to 91. Let x = the 1st number and (x+1) = the 2nd number, so that x + (x+1) =91 and 2x +1 =91, then 2x=90, x=45 and x+1=46. The answer is 45 and 46.

d. *Eliminate the possibilities and look for patterns or work the problem backwards*

e. *Guess the Answer*: Students should guess an approximate answer that makes sense based on the problem. For example, if the student knows the word problem implies addition, they should recognize that the answer must be greater than the numbers in the problem. Often students are afraid of guessing because they don't want to get the wrong answer but encourage your students to guess and then double check the answer to see if it works. If the answer is incorrect, the student can try another strategy for finding the answer.

f. *Make a Table*: Selecting relevant information from a word problem and organizing the data is very helpful in solving word problems. Often students become confused because there are too many numbers and/or variables.

http://www.purplemath.com/modules/translat.htm,
http://math.about.com/library/weekly/aa071002a.htm and
http://www.onlinemathlearning.com/algebra-word-problems.html are great resources for students to solve word problems.

6. Mental math practice

Give students regular practice in doing mental math. The following websites offer many mental calculation tips and strategies:
http://www.cramweb.com/math/index.htm
http://mathforum.org/k12/mathtips/mathtips.html

Because frequent calculator use tends to deprive students of a sense of numbers and an ability to calculate on their own, they will often approach a sequence of multiplications and divisions the hard way. For instance, asked to calculate 770 x 36/ 55, they will first multiply 770 and 36 and then do a long division with the 55. They fail to recognize that both 770 and 55 can be divided by 11 and then by 5 to considerably simplify the problem. Give students plenty of practice in multiplying and dividing a sequence of integers and fractions so they are comfortable with canceling top and bottom terms.

7. Math language
   Math vocabulary help is available for high school students on the web:

   http://www.amathsdictionaryforkids.com/ is a colorful website math
   dictionary
   http://www.math.com/tables/index.html is a math dictionary in English and
   Spanish

ERROR ANALYSIS

A simple method for analyzing student errors is to ask how the answer was
obtained. The teacher can then determine if a common error pattern has resulted
in the wrong answer. There is a value to having the students explain how they
arrived at the correct as well as the incorrect answers.

Many errors are due to simple **carelessness**. Students need to be encouraged
to work slowly and carefully. They should check their calculations by redoing the
problem on another paper, not merely looking at the work. Addition and
subtraction problems need to be written neatly so the numbers line up. Students
need to be careful regrouping in subtraction. Students must write clearly and
legibly, including erasing fully. Use estimation to ensure that answers make
sense.

Many students' computational skills exceed their **reading** level. Although they
can understand basic operations, they fail to grasp the concept or completely
understand the question. Students must read directions slowly.
Fractions are often a source of many errors. Students need to be reminded to
use common denominators when adding and subtracting and to always express
answers in simplest terms. Again, it is helpful to check by estimating.

The most common error that is made when working with **decimals** is failure to
line up the decimal points when adding or subtracting or not moving the decimal
point when multiplying or dividing. Students also need to be reminded to add
zeroes when necessary. Reading aloud may also be beneficial. Estimation, as
always, is especially important.

Students need to know that it is okay to make mistakes. The teacher must keep
a positive attitude, so they do not feel defeated or frustrated.

THE ART OF TEACHING - PEDAGOGICAL PRINCIPLES
Maintain a supportive, non-threatening environment

The key to success in teaching goes beyond your mathematical knowledge and the desire to teach. Though important, knowledge and desire alone do not make you a good teacher. Being able to connect with your students is vital: learn their names immediately, have a seating chart the first day (even if you intend to change it) and learn about your students -their hobbies, phone number, parent's names, what they like and dislike about school and learning and math. Keep this information on each student _and learn it:_ adapt your lessons, how challenging they are and what other resources you may need to accommodate your students' individual strengths and weaknesses.

Learn to see math as your students see it. If you aren't able to connect with your students, no matter how well your lessons are and how well you know the material, you won't inspire them to learn math from you. As you expect respect, you must give respect and as you expect their attention, they also need your attention and understanding. Talk to them with the same tone of voice as you would an adult, not in a tone that makes them feel like children. Look your students in the eye when you talk to them and encourage questions and comments. Take advantage of teachable moments and explain the rationale behind math rules.

Demonstrate respect, care and trust toward every student; assume the best. This does not mean becoming "friends" with your students or you will have problems with discipline. You can be kind and firm at the same time. Have a fair and clear grading and discipline system that is posted, reviewed and made clear to your students. Consistency, structure and fairness are essential to earning their trust in you as a teacher. Always admit your mistakes and be available to your students certain days after school. Finally, demonstrate your care for them and your love of math and you will be a positive influence on their learning.
Below are websites to help make your teaching more effective and fun:

1. **Teachers Helping Teachers** has several resources for high school mathematics.
2. **Math Resources for Teachers –** resources for grades 7 - 10
3. **Math is Marvelous Web Site** - is a fascinating website on the history of geometry
4. **Math Archives K-12** resources for lesson plans and software
5. **http://www.edhelper.com/algebra.htm** covers Algebra I & II
6. **Math Goodies** interactive lessons, worksheets and homework help
7. **Multicultural Lessons** an interesting site with lessons on Babylonian Square Roots, Chinese Fraction Reducing, Egyptian multiplication, etc.
8. **http://www.goenc.com/** resources and professional development for teachers

9. **Math and Reading Help** a guide to math, reading, homework help, tutoring and earning a high school diploma
10. **Purple Math.com** all about Algebra, lessons, help for students and lots of other resources
11. **http://library.thinkquest.org/20991/home.html** this site has Algebra, Geometry and Pre-calc/Calculus
12. **http://www.math.com/** this site has Algebra, Geometry, Trigonometry, and Calculus, plus homework help
13. **http://www.math.armstrong.edu/MathTutorial/index.html** a tutorial in algebra
14. **http://www.wtamu.edu/academic/anns/mps/math/mathlab/beg_algebr a/index.htm** this site is helpful for those beginning Algebra or for a refresher
15. **Math Complete** this radicals, quadratics, linear equation solvers
16. **Math Tutor - PEMDAS & Integers** fractions, integers, information for parents and teachers
17. **Matrix Algebra** all about matrix operations and applications
18. **Mr. Stroh's Algebra Site** help for Algebra I & II
19. **Polynomials and Polynomial Functions** everything from factoring, to graphing, finding rational zeros and multiplying, adding and subtracting polynomials
20. **Quadratic Formula** all about quadratics
21. **Animated Pythagorean Theorem** a fun an animated proof of the Pythagorean Theorem
22. **Brunnermath - Interactive Activities** general math, Algebra, Geometry, Trigonometry, Statistics, Calculus, using Calculators
23. **CoolMath4Kids Geometry** creating art with math and geometry lessons
24. **The Curlicue Fractal** The curlicue fractal is an exceedingly easy-to-make but richly complex pattern using trigonometry and calculus to create fascinating shapes
25. **Euclid's Elements Interactive** Euclid's *Elements* form one of the most beautiful and influential works of science in the history of humankind.
26. **Howe-Two Free Software** software solutions for mathematics instruction
27. **http://regentsprep.org/regents/math/math-topic.cfm?TopicCode=syslin** Systems of equations lessons and practice
28. **http://www.sparknotes.com/math/algebra1/systemsofequations/probl ems3.rhtml** Word problems system of equations
29. **http://math.about.com/od/complexnumbers/Complex_Numbers.htm** Several complex number exercise pages
30. **http://regentsprep.org/Regents/math/ALGEBRA/AE3/PracWord.htm** practice with Systems of inequalities word problems
31. **http://regentsprep.org/regents/Math/solvin/PSolvln.htm** solving inequalities
32. **http://www.wtamu.edu/academic/anns/mps/math/mathlab/beg_algebr a/beg_alg_tut18_ineq.htm** Inequality tutorial, examples, problems

33. **http://www.wtamu.edu/academic/anns/mps/math/mathlab/beg_algebr a/beg_alg_tut24_ineq.htm** Graphing linear inequalities tutorial
34. **http://www.wtamu.edu/academic/anns/mps/math/mathlab/col_algebr a/col_alg_tut17_quad.htm** Quadratic equations tutorial, examples, problems
35. **http://regentsprep.org/Regents/math/math- topic.cfm?TopicCode=factor** Practice factoring
36. **http://www.wtamu.edu/academic/anns/mps/math/mathlab/col_algebr a/col_alg_tut37_syndiv.htm** Synthetic division tutorial
37. **http://www.tpub.com/math1/10h.htm** Synthetic division Examples and problems

DEVELOPMENTAL PSYCHOLOGY AND TEACHING MATHEMATICS- things you may not know about your students:

Studies show that health matters more than gender or social status when it comes to learning. Healthy girls and boys do equally well on most cognitive tasks. Boys perform better at manipulating shapes and analyzing and girls perform better on processing speed and motor dexterity. No differences have been measured in calculation ability, meaning girls and boys have an equal aptitude for mathematics.

The following was written by Jay Giedd, M.D. is a practicing Child and Adolescent Psychiatrist and Chief of Brain Imaging at the Child Psychiatry Branch of the National Institute of Mental Health:

http://nihrecord.od.nih.gov/newsletters/2005/08_12_2005/story04.htm

"The most surprising thing has been how much the teen brain is changing. By age six, the brain is already 95 percent of its adult size. But the gray matter, or thinking part of the brain, continues to thicken throughout childhood as the brain cells get extra connections, much like a tree growing extra branches, twigs and roots...In the frontal part of the brain, the part of the brain involved in judgment, organization, planning, strategizing -- those very skills that teens get better and better at -- this process of thickening of the gray matter peaks at about age 11 in girls and age 12 in boys, roughly about the same time as puberty. After that peak, the gray matter thins as the excess connections are eliminated or pruned...

Contrary to what most parents have thought at least once, "teens really do have brains," quipped Dr. Jay Giedd, NIMH intramural scientist, in a lecture on the "Teen Brain Under Construction." His talk was the kick-off event for the recent NIH Parenting Festival. Giedd said scientists have only recently learned more about the trajectories of brain growth. One of the findings he discussed showed the frontal cortex area — which governs judgment, decision-making and impulse control — doesn't fully mature until around age 25. "That really threw us," he said. "We used to joke about having to be 25 to rent a car, but there's tons of data from insurance reports [showing] that 24-year-olds are costing them more than 44-year-olds."

So why is that? "It must be behavior and impulse control," he said. "Whatever these changes are, the top 10 bad things that happen to teens involve emotion and behavior." Physically, Giedd said, the teen years and early 20s represent an incredibly healthy time of life, in terms of cancer, heart disease and other serious illnesses. But with accidents as the leading cause of death in adolescents, and suicide following close behind, "this isn't a great time emotionally and psychologically. This is the great paradox of adolescence: right at the time you should be on the top of your game, you're not."

The next step in Giedd's research, he said, is to learn more about what influences brain growth, for good or bad. "Ultimately, we want to use these findings to treat illness and enhance development."

One of the things scientists have come to understand, though, is that parents do have something to do with their children's brain development.

"From imaging studies, one of the things that seems intriguing is this notion of modeling...that the brain is pretty adept at learning by example," he said. "As parents, we teach a lot when we don't even know we're teaching, just by showing how we treat our spouses, how we treat other people, what we talk about in the car on the way home...things that a parent says in the car can stick with them for years. They're listening even though it may appear they're not."

What can we do to change our kids? "Well, start with yourself in terms of what you show by example," Giedd concluded.

Maybe the parts of the brain performing geometry are different from the parts doing algebra. There is no definitive research to answer that question yet, but it is obviously what researchers are looking for.

Time-Lapse Imaging Tracks Brain Maturation from ages 5 to 20
Constructed from MRI scans of healthy children and teens, the time-lapse "movie", from which the above images were extracted, compresses 15 years of brain development (ages 5–20) into just a few seconds.

To view in color, go to the website below: Red indicates **more** gray matter, **blue less** gray matter. Gray matter wanes in a back-to-front wave as the brain matures and neural connections are pruned.

Source: Paul Thompson, Ph.D. UCLA Laboratory of Neuroimaging
http://www.loni.ucla.edu/%7Ethompson/DEVEL/PR.html

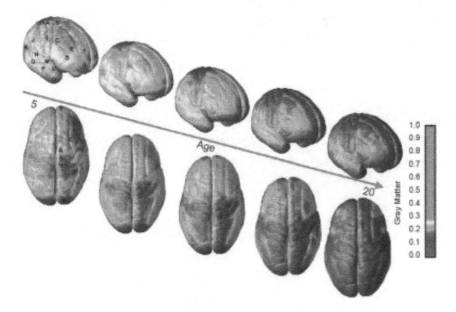

What are the implications of this fascinating study for teachers? It's unreasonable to expect teens to have adult levels of organizational skills or decision-making before their brains have completely developed. In teens, the frontal lobe, or the executive of the brain is what handles organizing, decision making, emotions, attending, shifting attention, planning and making strategies and it is not fully developed until the early to mid-twenties.

*Perhaps since certain parts of the brain develop sooner than others, subjects should be taught in a different order. Until we know more, just understanding that the parts of teen brains related to decision making and emotions are still developing through the early 20's is important, and **that stressful situations lead to diminished ability to made good judgments.** For some children, just being called on in class is stressful. At this age, social relationships become very important and **teachers need to be sensitive to this aspect of teen development as it relates to stress and decision-making.** The immaturity of this part of the teen brain might explain why the teen crash rate is **4** times that of adults…*

ANSWER KEY TO PRACTICE PROBLEMS

<u>0001</u>

Page 5

Question #1   The Red Sox won the World Series.
Question #2   Angle B is not between 0 and 90 degrees.
Question #3   Annie will do well in college.
Question #4   You are witty and charming.

<u>0005</u>

Page 38

Question #1 $(6x - 5y)(36x^2 + 30xy + 25y^2)$

Question #2 $4(a - 2b)(a^2 + 2ab + 4b^2)$

Question #3 $5x^2(2x^9 + 3y)(4x^{18} - 6x^9 y + 9y^2)$

Page 39

Question #1 $(2x - 5y)(2x + 5y)$

Question #2 $2(3b - 4)(b + 1)$

Question #3   The answer is D   (2x+1)

Page 40

Question #2   a, b, c, f are functions
Question #3   Domain = $^-\infty, \infty$ Range = $^-5, \infty$

Page 42

Question #1   Domain = $^-\infty, \infty$ Range = $^-6, \infty$
Question #2   Domain = 1,4,7,6   Range = -2
Question #3   Domain = $x \neq 2, ^-2$
Question #4   Domain = $^-\infty, \infty$ Range = -4, 4

                    Domain = $^-\infty, \infty$ Range = $2, \infty$

Question #5   Domain = $^-\infty, \infty$ Range = 5
Question #6   (3,9), (-4,16), (6,3), (1,9), (1,3)

<u>0006</u>

Page 44

Question #1

Question #2

Question #3

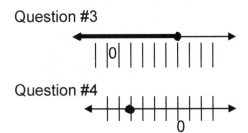

Question #4

Page 46

Question #1

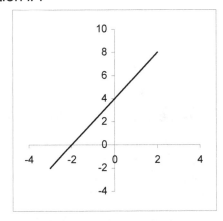

Question #2

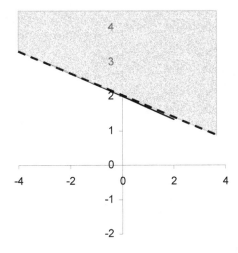

Question #3

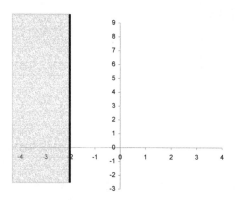

Page 48

Question #1   x-intercept = -14   y-intercept = -10   slope = $-\dfrac{5}{7}$

Question #2   x-intercept = 14   y-intercept = -7   slope = $\dfrac{1}{2}$

Question #3 x-intercept = 3   y-intercept = none

Question #4 x-intercept = $\dfrac{15}{2}$ y-intercept = 3   slope = $-\dfrac{2}{5}$

Page 48

Question #1 $y = \dfrac{3}{4}x + \dfrac{17}{4}$

Question #2 $x = 11$

Question #3 $y = \dfrac{3}{5}x + \dfrac{42}{5}$

Question #4 $y = 5$

Page 50

Question #1 $\begin{pmatrix} x \\ y \end{pmatrix} = \begin{pmatrix} 3 \\ 1 \end{pmatrix}$

Question #2 $\begin{pmatrix} x \\ y \\ z \end{pmatrix} = \begin{pmatrix} 4 \\ 4 \\ 1 \end{pmatrix}$

0007

Page 65

Question #1   The sides are 8, 15, and 17

Question #2   The numbers are 2 and $\dfrac{1}{2}$

Page 66

Question #1 $x = \dfrac{7 \pm \sqrt{241}}{12}$

Question #2 $x = \dfrac{1}{2}$ or $\dfrac{^-2}{5}$

Question #3 $x = \dfrac{8}{5}$

Page 68

Question #1 $x^2 - 10x + 25$

Question #2 $25x^2 - 10x - 48$

Question #3 $x^2 - 9x - 36$

<u>0008</u>

Page 82

Question #1

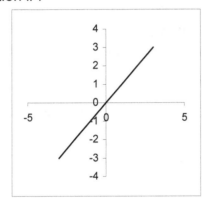

Question #2

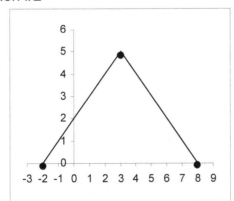

Question #3

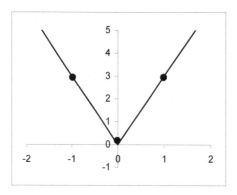

Question #4

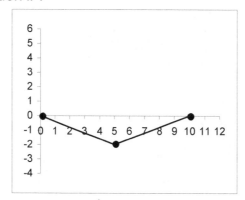

0009

Page 86

Question #2 $x = 9$
Question #3 $x = 3$ or $x = 6$
Question #4   1

Page 87

Question #1 $\dfrac{^-12x^5z^5}{y}$

Question #2 $25a^4 + 15a^2 - 7b^3$
Question #3 $100x^{20}y^8 + 8x^9y^6$

Page 94

Question #1 $\quad \cot\theta = \dfrac{x}{y}$

$$\frac{x}{y} = \frac{x}{r} \times \frac{r}{y} = \frac{x}{y} = \cot\theta$$

Question #2 $\quad 1 + \cot^2\theta = \csc^2\theta$

$$\frac{y^2}{y^2} + \frac{x^2}{y^2} = \frac{r^2}{y^2} = \csc^2\theta$$

0012

Page 117

Question #1   49.34
Question #2   1

Page 118

Question #1 ∞
Question #2   -1

Page 129

Question #1 $\quad S_5 = 75$

Question #2 $\quad S_n = 28$

Question #3 $\quad S_n = -\dfrac{-31122}{15625} \approx^{-} 1.99$

0013

Page 139

Question #1 $\quad t(0) = -24$ m/sec
Question #2 $\quad t(4) = 24$ m/sec

Page 155

Question #1 $\quad g(x) = -20\sin x + c$
Question #2 $\quad g(x) = \pi\sec x + c$

0018

Page 211

Question #1 $\quad 35x^3 y^4$

Question #2 $\quad 78732x^7 y^2$

SAMPLE TEST

Directions: Read each item and select the best response.

1. **When you begin by assuming the conclusion of a theorem is false, then show that through a sequence of logically correct steps you contradict an accepted fact, this is known as (Easy)(Skill 0001)**

   A) inductive reasoning
   B) direct proof
   C) indirect proof
   D) exhaustive proof

2. **Given the series of examples below, what is 5¢4?**

   $4\not\subset 3=13$ $7\not\subset 2=47$
   $3\not\subset 1=8$ $1\not\subset 5=-4$

   (Average Rigor)(Skill 0001)

   A) 20
   B) 29
   C) 1
   D) 21

3. **A group of students working with trigonometric identities have concluded that $\cos 2x = 2\cos x$. How could you best lead them to discover their error?**
   (Average Rigor)(Skill 0001)

   A) Have the students plug in values on their calculators.
   B) Direct the student to the appropriate chapter in the text.
   C) Derive the correct identity on the board.
   D) Provide each student with a table of trig identities.

4. **Which best describes the type of error observed below?**

-3 + 7 = -4	6(-10) = - 60
-5(-15) = 75	-3+-8 = 11
8-12 = -4	7- -8 = 15

   (Easy)(Skill 0001)

   A) The student is incorrectly multiplying integers.
   B) The student has incorrectly applied rules for adding integers to subtracting integers.
   C) The student has incorrectly applied rules for multiplying integers to adding integers.
   D) The student is incorrectly subtracting integers.

5. About two weeks after introducing formal proofs, several students in your geometry class are having a difficult time remembering the names of the postulates. They cannot complete the reason column of the proof and as a result are not even attempting the proofs. What would be the best approach to help students understand the nature of geometric proofs?
(Average Rigor)(Skill 0001)

   A) Give them more time; proofs require time and experience.
   B) Allow students to write an explanation of the theorem in the reason column instead of the name.
   C) Have the student copy each theorem in a notebook.
   D) Allow the students to have open book tests.

6. Which mathematician is best known for his work in developing non-Euclidean geometry?
(Easy)(Skill 0002)

   A) Descartes
   B) Riemann
   C) Pascal
   D) Pythagoras

7. According to Piaget, what stage in a student's development would be appropriate for introducing abstract concepts in geometry?
(Easy)(Skill 0002)

   A) concrete operational
   B) formal operational
   C) sensori-motor
   D) pre-operational

8. Change $.\overline{63}$ into a fraction in simplest form.
(Easy)(Skill 0002)

   A) 63/100
   B) 7/11
   C) 6 3/10
   D) 2/3

9. A student had 60 days to appeal the results of an exam. If the results were received on March 23, what was the last day that the student could appeal?
(Average Rigor)(Skill 0002)

   A) May 21
   B) May 22
   C) May 23
   D) May 24

10. Express .0000456 in scientific notation.
(Easy)(Skill 0002)

   A) $4.56x10^{-4}$
   B) $45.6x10^{-6}$
   C) $4.56x10^{-6}$
   D) $4.56x10^{-5}$

11. The volume of water flowing through a pipe varies directly with the square of the radius of the pipe. If the water flows at a rate of 80 liters per minute through a pipe with a radius of 4 cm, at what rate would water flow through a pipe with a radius of 3 cm?
(Rigorous)(Skill 0003)

   A) 45 liters per minute
   B) 6.67 liters per minute
   C) 60 liters per minute
   D) 4.5 liters per minute

12. The mass of a Chips Ahoy cookie would be to
(Average Rigor)(Skill 0003)

   A) 1 kilogram
   B) 1 gram
   C) 15 grams
   D) 15 milligrams

13. What would be the least appropriate use for handheld calculators in the classroom?
(Average Rigor)(Skill 0003)

   A) practice for standardized tests
   B) integrating algebra and geometry with applications
   C) justifying statements in geometric proofs
   D) applying the law of sines to find dimensions

14. What would be the total cost of a suit for $295.99 and a pair of shoes for $69.95 including 6.5% sales tax?
(Average Rigor)(Skill 0003)

   A) $389.73
   B) $398.37
   C) $237.86
   D) $315.23

15. Which of the following is the best example of the value of personal computers in advanced high school mathematics?
(Easy)(Skill 0003)

    A) Students can independently drill and practice test questions.
    B) Students can keep an organized list of theorems and postulates on a word processing program.
    C) Students can graph and calculate complex functions to explore their nature and make conjectures.
    D) Students are better prepared for business because of mathematics computer programs in high school.

16. Which of the following illustrates an inverse property?
(Easy)(Skill 0004)

    A) $a + b = a - b$
    B) $a + b = b + a$
    C) $a + 0 = a$
    D) $a + (-a) = 0$

17. Simplify: $\sqrt{27} + \sqrt{75}$
(Rigorous)(Skill 0004)

    A) $8\sqrt{3}$
    B) 34
    C) $34\sqrt{3}$
    D) $15\sqrt{3}$

18. Determine the number of subsets of set K.

    $K = \{4, 5, 6, 7\}$

(Average Rigor)(Skill 0004)

    A) 15
    B) 16
    C) 17
    D) 18

19. Simplify: $\dfrac{10}{1+3i}$
(Average Rigor)(Skill 0004)

    A) $-1.25(1-3i)$
    B) $1.25(1+3i)$
    C) $1+3i$
    D) $1-3i$

20. Which of the following sets is closed under division?
(Average Rigor)(Skill 0004)

    I) $\{½, 1, 2, 4\}$
    II) $\{-1, 1\}$
    III) $\{-1, 0, 1\}$

    A) I only
    B) II only
    C) III only
    D) I and II

21. Identify the correct sequence of subskills required for solving and graphing inequalities involving absolute value in one variable, such as $|x+1| \leq 6$.
(Average Rigor)(Skill 0005)

   A) understanding absolute value, graphing inequalities, solving systems of equations
   B) graphing inequalities on a Cartesian plane, solving systems of equations, simplifying expressions with absolute value
   C) plotting points, graphing equations, graphing inequalities
   D) solving equations with absolute value, solving inequalities, graphing conjunctions and disjunctions

22. Which of the following is always composite if x is odd, y is even, and both x and y are greater than or equal to 2?
(Average Rigor)(Skill 0005)

   A) $x+y$
   B) $3x+2y$
   C) $5xy$
   D) $5x+3y$

23. Solve for v_0 : $d = at(v_t - v_0)$
(Average Rigor)(Skill 0005)

   A) $v_0 = atd - v_t$
   B) $v_0 = d - atv_t$
   C) $v_0 = atv_t - d$
   D) $v_0 = (atv_t - d)/at$

24. $f(x) = 3x - 2;\ f^{-1}(x) =$
(Average Rigor)(Skill 0006)

   A) $3x + 2$
   B) $x/6$
   C) $2x - 3$
   D) $(x+2)/3$

25. State the domain of the function $f(x) = \dfrac{3x-6}{x^2-25}$
(Average Rigor)(Skill 0006)

   A) $x \neq 2$
   B) $x \neq 5, -5$
   C) $x \neq 2, -2$
   D) $x \neq 5$

26. What is the equation of the graph below?
(Easy)(Skill 0006)

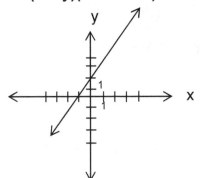

A) $2x + y = 2$
B) $2x - y = -2$
C) $2x - y = 2$
D) $2x + y = -2$

27. Which of the following is a factor of $6 + 48m^3$
(Rigorous)(Skill 0007)

A) (1 + 2m)
B) (1 - 8m)
C) (1 + m - 2m)
D) (1 - m + 2m)

28. Given $f(x) = 3x - 2$ and $g(x) = x^2$, determine $g(f(x))$.
(Average Rigor)(Skill 0007)

A) $3x^2 - 2$
B) $9x^2 + 4$
C) $9x^2 - 12x + 4$
D) $3x^3 - 2$

29. Which graph represents the equation of $y = x^2 + 3x$?
(Average Rigor)(Skill 0007)

A)

B)

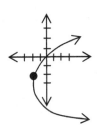

C)

D)

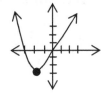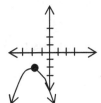

30. Which of the following is incorrect?
(Rigorous)(Skill 0007)

A) $(x^2 y^3)^2 = x^4 y^6$
B) $m^2 (2n)^3 = 8m^2 n^3$
C) $(m^3 n^4)/(m^2 n^2) = mn^2$
D) $(x + y^2)^2 = x^2 + y^4$

31. Which graph represents the solution set for $x^2 - 5x > -6$?
(Average Rigor)(Skill 0007)

A)
    -2   0   2

B)
    -3   0

C)
    -2   0   2

D)
    -3   0   2 3

32. What would be the seventh term of the expanded binomial $(2a+b)^8$?
(Rigorous)(Skill 0007)

A) $2ab^7$
B) $41a^4b^4$
C) $112a^2b^6$
D) $16ab^7$

33. Find the zeroes of
$f(x) = x^3 + x^2 - 14x - 24$
(Rigorous)(Skill 0007)

A) 4, 3, 2
B) 3, -8
C) 7, -2, -1
D) 4, -3, -2

34. Evaluate $3^{1/2}(9^{1/3})$
(Rigorous)(Skill 0008)

A) $27^{5/6}$
B) $9^{7/12}$
C) $3^{5/6}$
D) $3^{6/7}$

35. Solve for x: $18 = 4 + |2x|$
(Rigorous)(Skill 0008)

A) $\{-11, 7\}$
B) $\{-7, 0, 7\}$
C) $\{-7, 7\}$
D) $\{-11, 11\}$

36. Which equation corresponds to the logarithmic statement: $\log_x k = m$?
(Rigorous)(Skill 0009)
A) $x^m = k$
B) $k^m = x$
C) $x^k = m$
D) $m^x = k$

37. Which of the following is the best approximate value of x in the following equation?

$$\ln(x) = 8$$

(Easy)(Skill 0009)

A) 2981
B) -2981
C) 2.079
D) -2.079

38. Solve for x in the following equation:

$$a \log(3x) = b$$

(Average Rigor)(Skill 0009)

A) $x = 30^{\frac{b}{a}}$

B) $x = \dfrac{10^{\frac{a}{b}}}{3}$

C) $x = \left(\dfrac{10}{3}\right)^{\frac{b}{a}}$

D) $x = \dfrac{10^{\frac{b}{a}}}{3}$

39. The populations of two towns are growing exponentially as functions of time:

Town 1: $P(t) = 100e^{0.010t}$

Town 2: $P(t) = 120e^{0.008t}$

Where t is positive and given in years. It t=0 in the year 2000, in what year will the towns have the same population?

(Rigorous)(Skill 0009)

A) 2044
B) 2091
C) 2910
D) 2050

40. Which expression is equivalent to $1 - \sin^2 x$?
(Rigorous)(Skill 0010)

A) $1 - \cos^2 x$
B) $1 + \cos^2 x$
C) $1/\sec x$
D) $1/\sec^2 x$

41. Find the slope of the line tangent to $y = 3x(\cos x)$ at $(\pi/2,\ \pi/2)$.
(Rigorous)(Skill 0010)

A) $-3\pi/2$
B) $3\pi/2$
C) $\pi/2$
D) $-\pi/2$

42. Find the equation of the line tangent to $y = 3x^2 - 5x$ at $(1, -2)$.
(Rigorous)(Skill 0010)

A) $y = x - 3$
B) $y = 1$
C) $y = x + 2$
D) $y = x$

43. Determine the measures of angles A and B.
(Average Rigor)(Skill 0011)

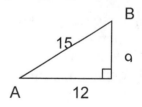

A) $A = 30°,\quad B = 60°$
B) $A = 60°,\quad B = 30°$
C) $A = 53°,\quad B = 37°$
D) $A = 37°,\quad B = 53°$

44. What is the measure of minor arc AD, given measure of arc PS is 40° and $m < K = 10°$?
(Rigorous)(Skill 0011)

A) 50°
B) 20°
C) 30°
D) 25°

45. Find the first derivative of the function: $f(x) = x^3 - 6x^2 + 5x + 4$
(Rigorous)(Skill 0012)

A) $3x^3 - 12x^2 + 5x = f'(x)$
B) $3x^2 - 12x - 5 = f'(x)$
C) $3x^2 - 12x + 9 = f'(x)$
D) $3x^2 - 12x + 5 = f'(x)$

46. Differentiate: $y = e^{3x+2}$
(Rigorous)(Skill 0012)

A) $3e^{3x+2} = y'$
B) $3e^{3x} = y'$
C) $6e^3 = y'$
D) $(3x + 2)e^{3x+1} = y'$

47. How does the function $y = x^3 + x^2 + 4$ behave from $x = 1$ to $x = 3$?
(Average Rigor)(Skill 0012)

A) increasing, then decreasing
B) increasing
C) decreasing
D) neither increasing nor decreasing

48. Find the absolute maximum obtained by the function $y = 2x^2 + 3x$ on the interval $x = 0$ to $x = 3$.
(Rigorous)(Skill 0012)

A) $-3/4$
B) $-4/3$
C) 0
D) 27

49. Find the antiderivative for $4x^3 - 2x + 6 = y$.
(Rigorous)(Skill 0012)

A) $x^4 - x^2 + 6x + C$
B) $x^4 - 2/3x^3 + 6x + C$
C) $12x^2 - 2 + C$
D) $4/3x^4 - x^2 + 6x + C$

50. Evaluate: $\int (x^3 + 4x - 5)dx$
(Rigorous)(Skill 0012)

A) $3x^2 + 4 + C$
B) $\frac{1}{4}x^4 - 2/3x^3 + 6x + C$
C) $x^{4/3} + 4x - 5x + C$
D) $x^3 + 4x^2 - 5x + C$

51. Evaluate $\int_0^2 (x^2 + x - 1)dx$
(Rigorous)(Skill 0012)

A) 11/3
B) 8/3
C) -8/3
D) -11/3

52. Find the area under the function $y = x^2 + 4$ from $x = 3$ to $x = 6$.
(Average Rigor)(Skill 0012)

A) 75
B) 21
C) 96
D) 57

53. Find the antiderivative for the function $y = e^{3x}$.
(Rigorous)(Skill 0012)

A) $3x(e^{3x}) + C$
B) $3(e^{3x}) + C$
C) $1/3(e^x) + C$
D) $1/3(e^{3x}) + C$

54. The acceleration of a particle is dv/dt = 6 m/s². Find the velocity at t=10 given an initial velocity of 15 m/s.
(Average Rigor)(Skill 0013)

A) 60 m/s
B) 150 m/s
C) 75 m/s
D) 90 m/s

55. If the velocity of a body is given by v = 16 - t², find the distance traveled from t = 0 until the body comes to a complete stop.
(Average Rigor)(Skill 0013)

A) 16
B) 43
C) 48
D) 64

56. Compute the area of the shaded region, given a radius of 5 meters. 0 is the center.
(Rigorous)(Skill 0014)

A) 7.13 cm²
B) 7.13 m²
C) 78.5 m²
D) 19.63 m²

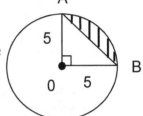

57. If the area of the base of a cone is tripled, the volume will be
(Rigorous)(Skill 0014)

A) the same as the original
B) 9 times the original
C) 3 times the original
D) 3 π times the original

58. Choose the correct statement concerning the median and altitude in a triangle.
(Average Rigor)(Skill 0014)

A) The median and altitude of a triangle may be the same segment.
B) The median and altitude of a triangle are always different segments.
C) The median and altitude of a right triangle are always the same segment.
D) The median and altitude of an isosceles triangle are always the same segment.

59. Find the surface area of a box which is 3 feet wide, 5 feet tall, and 4 feet deep.
    (Easy)(Skill 0014)

    A) 47 sq. ft.
    B) 60 sq. ft.
    C) 94 sq. ft
    D) 188 sq. ft.

60. Given a 30 meter x 60 meter garden with a circular fountain with a 5 meter radius, calculate the area of the portion of the garden not occupied by the fountain.
    (Average Rigor)(Skill 0014)

    A) 1721 m²
    B) 1879 m²
    C) 2585 m²
    D) 1015 m²

61. Determine the area of the shaded region of the trapezoid in terms of x and y.
    (Rigorous)(Skill 0014)

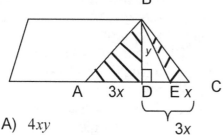

    A) $4xy$
    B) $2xy$
    C) $3x^2y$
    D) There is not enough information given.

62. Find the area of the figure pictured below.
    (Rigorous)(Skill 0014)

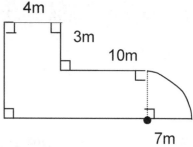

    A) 136.47 m²
    B) 148.48 m²
    C) 293.86 m²
    D) 178.47 m²

63. Choose the diagram which illustrates the construction of a perpendicular to the line at a given point on the line.
    (Rigorous)(Skill 0015)

A)

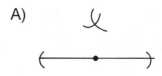

B)

C)

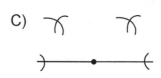

D)

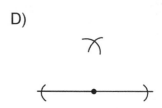

64. Which term most accurately describes two coplanar lines without any common points?
(Easy)(Skill 0015)

   A) perpendicular
   B) parallel
   C) intersecting
   D) skew

65. What is the degree measure of an interior angle of a regular 10 sided polygon?
(Rigorous)(Skill 0015)

   A) 18°
   B) 36°
   C) 144°
   D) 54°

66. Which theorem can be used to prove $\triangle BAK \cong \triangle MKA$?
(Average Rigor)(Skill 0015)

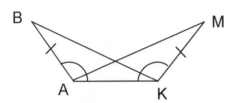

   A) SSS
   B) ASA
   C) SAS
   D) AAS

67. Find the length of the major axis of $x^2 + 9y^2 = 36$.
(Rigorous)(Skill 0015)
   A) 4
   B) 6
   C) 12
   D) 8

68. Which equation represents a circle with a diameter whose endpoints are $(0,7)$ and $(0,3)$?
(Rigorous)(Skill 0016)
   A) $x^2 + y^2 + 21 = 0$
   B) $x^2 + y^2 - 10y + 21 = 0$
   C) $x^2 + y^2 - 10y + 9 = 0$
   D) $x^2 - y^2 - 10y + 9 = 0$

69. Given that QO⊥NP and QO=NP, quadrilateral NOPQ can most accurately be described as a
(Easy)(Skill 0016)

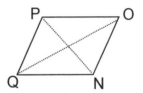

   A) parallelogram
   B) rectangle
   C) square
   D) rhombus

70. Given $K(-4, y)$ and $M(2,-3)$ with midpoint $L(x,1)$, determine the values of x and y.
(Rigorous)(Skill 0016)

   A) $x = -1,\ y = 5$
   B) $x = 3,\ y = 2$
   C) $x = 5,\ y = -1$
   D) $x = -1,\ y = -1$

71. Determine the rectangular coordinates of the point with polar coordinates (5, 60°). (Average Rigor)(Skill 0016)

    A) (0.5, 0.87)
    B) (-0.5, 0.87)
    C) (2.5, 4.33)
    D) (25, 150°)

72. Given a vector with horizontal component 5 and vertical component 6, determine the length of the vector. (Average Rigor)(Skill 0017)

    A) 61
    B) $\sqrt{61}$
    C) 30
    D) $\sqrt{30}$

73. Compute the distance from (-2,7) to the line $x = 5$. (Average Rigor)(Skill 0017)
    A) -9
    B) -7
    C) 5
    D) 7

74. If a ship sails due south 6 miles, then due west 8 miles, how far was it from the starting point? (Average Rigor)(Skill 0017)

    A) 100 miles
    B) 10 miles
    C) 14 miles
    D) 48 miles

75. Find the sum of the first one hundred terms in the progression.
    (-6, -2, 2 . . .)
(Rigorous)(Skill 0018)

    A) 19,200
    B) 19,400
    C) -604
    D) 604

76. What is the value of the following expression?

$$\frac{8!}{3!}$$

(Easy)(Skill 0018)

    A) 13440
    B) 120
    C) 6720
    D) 4.01

77. How many ways are there to choose a potato and two green vegetables from a choice of three potatoes and seven green vegetables? (Average Rigor)(Skill 00018)

    A) 126
    B) 63
    C) 21
    D) 252

78. A jar contains 6 red beans, 5 black beans, and 8 garbanzo beans. If a single bean is chosen at random, what the probability of it being a red bean?

(Average Rigor)(Skill 0019)

A) 18.6%
B) 25.6%
C) 52.3%
D) 31.5%

79. A jar contains 6 red beans, 5 black beans, and 8 garbanzo beans. If two beans are drawn in sequence, what is the probability that first a red bean and then a black bean will be drawn (assume the red bean is not replaced after being drawn)?

(Rigorous)(Skill 0019)

A) 88.2%
B) 27.7%
C) 8.7%
D) 8.2%

80. A jar contains 6 red beans, 5 black beans, and 8 garbanzo beans. Given that a red bean has been drawn (and not replaced), what is the probability that the next bean drawn will be black?

(Rigorous)(Skill 0019)

A) 31.5%
B) 8.7%
C) 25.6%
D) 27.7%

81. The probability distribution shown below exhibits:
(Average Rigor)(Skill 0019)

A) positive skew
B) negative skew
C) excessive kurtosis
D) diminished kurtosis

82. The number of pizza slices eaten per college student per year fits a normal distribution with a mean of 55 and a standard deviation of 15. The number of pizza slices eaten annually by the students in the top 2.5% of the distribution is greater than:
(Rigorous)(Skill 0019)

A) 70
B) 85
C) 100
D) 110

83. Compute the median for the following data set:
{12, 19, 13, 16, 17, 14}
(Easy)(Skill 0020)

A) 14.5
B) 15.17
C) 15
D) 16

84. Compute the standard deviation for the following set of temperatures.
(37, 38, 35, 37, 38, 40, 36, 39)
(Easy)(Skill 0020)

   A) 37.5
   B) 1.5
   C) 0.5
   D) 2.5

85. Half the students in a class scored 80% on an exam, most of the rest scored 85% except for one student who scored 10%. Which would be the best measure of central tendency for the test scores?
(Rigorous)(Skill 0020)

   A) mean
   B) median
   C) mode
   D) either the median or the mode because they are equal

86. What conclusion can be drawn from the graph below?

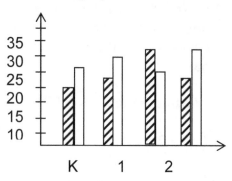

MLK Elementary Student Enrollment
(Easy)(Skill 0021)

Girls Boys

   A) The number of students in first grade exceeds the number in second grade.
   B) There are more boys than girls in the entire school.
   C) There are more girls than boys in the first grade.
   D) Third grade has the largest number of students.

87. Solve the system of equations for x, y and z.

$$3x + 2y - z = 0$$
$$2x + 5y = 8z$$
$$x + 3y + 2z = 7$$

(Rigorous)(Skill 0021)

   A) $(-1, 2, 1)$
   B) $(1, 2, -1)$
   C) $(-3, 4, -1)$
   D) $(0, 1, 2)$

88. Evaluate the dot product of the following matrices as shown:

$$\begin{vmatrix} 1 & 5 \\ 7 & 3 \end{vmatrix} \bullet \begin{vmatrix} 4 & 5 \\ 2 & 9 \end{vmatrix}$$

(Average Rigor)(Skill 0021)

A) 70

B)
$$\begin{vmatrix} 4 & 50 \\ 34 & 67 \end{vmatrix}$$

C) 91

D)
$$\begin{vmatrix} 4 & 50 \\ 34 & 67 \end{vmatrix}$$

89. Find the value of the determinant of the matrix.
(Average Rigor)(Skill 0021)

$$\begin{vmatrix} 2 & 1 & -1 \\ 4 & -1 & 4 \\ 0 & -3 & 2 \end{vmatrix}$$

A) 0
B) 23
C) 24
D) 40

90. What would be the shortest method of solution for the system of equations below?
$$3x + 2y = 38$$
$$4x + 8 = y$$
(Easy)(Skill 0021)

A) linear combination
B) additive inverse
C) substitution
D) graphing

Constructed Response Problem:

The town of Verdant Slopes has been experiencing a boom in population growth. By the year 2000, the population had grown to 45,000, and by 2005, the population had reached 60,000.

a. Using the formula for slope as a model, find the average rate of change in population growth, expressing your answer in people per year.

b. Using the average rate of change determined in a., predict the population of Verdant Slopes in the year 2010.

Answer Key

1.	C	24.	D	47.	B	70.	A
2.	D	25.	B	48.	D	71.	C
3.	C	26.	B	49.	A	72.	B
4.	C	27.	A	50.	B	73.	D
5.	B	28.	C	51.	B	74.	B
6.	B	29.	C	52.	A	75.	C
7.	A	30.	D	53.	D	76.	B
8.	B	31.	D	54.	C	77.	A
9.	B	32.	C	55.	B	78.	D
10.	D	33.	D	56.	B	79.	C
11.	A	34.	B	57.	C	80.	D
12.	C	35.	C	58.	A	81.	A
13.	C	36.	A	59.	C	82.	B
14.	A	37.	A	60.	A	83.	C
15.	C	38.	D	61.	B	84.	B
16.	D	39.	B	62.	B	85.	B
17.	A	40.	D	63.	D	86.	B
18.	B	41.	A	64.	B	87.	A
19.	D	42.	A	65.	C	88.	A
20.	B	43.	D	66.	C	89.	C
21.	D	44.	B	67.	C	90.	C
22.	C	45.	D	68.	B		
23.	D	46.	A	69.	C		

Rigor Analysis Table

Questions	Assessments
1	Easy
2	Average Rigor
3	Average Rigor
4	Easy
5	Average Rigor
6	Easy
7	Easy
8	Easy
9	Average Rigor
10	Easy
11	Rigorous
12	Average Rigor
13	Average Rigor
14	Average Rigor
15	Easy
16	Easy
17	Rigorous
18	Average Rigor
19	Average Rigor
20	Average Rigor
21	Average Rigor
22	Average Rigor
23	Average Rigor
24	Average Rigor
25	Average Rigor
26	Easy
27	Rigorous
28	Average Rigor
29	Average Rigor
30	Rigorous
31	Average Rigor
32	Rigorous
33	Rigorous
34	Rigorous
35	Rigorous
36	Rigorous
37	Easy
38	Average Rigor
39	Rigorous
40	Rigorous
41	Rigorous

42	Rigorous
43	Average Rigor
44	Rigorous
45	Rigorous
46	Rigorous
47	Average Rigor
48	Rigorous
49	Rigorous
50	Rigorous
51	Rigorous
52	Average Rigor
53	Rigorous
54	Average Rigor
55	Average Rigor
56	Rigorous
57	Rigorous
58	Average Rigor
59	Easy
60	Average Rigor
61	Rigorous
62	Rigorous
63	Rigorous
64	Easy
65	Rigorous
66	Average Rigor
67	Rigorous
68	Rigorous
69	Easy
70	Rigorous
71	Average Rigor
72	Average Rigor
73	Average Rigor
74	Average Rigor
75	Rigorous
76	Easy
77	Average Rigor
78	Average Rigor
79	Rigorous
80	Rigorous
81	Average Rigor
82	Rigorous
83	Easy
84	Easy
85	Rigorous

86	Easy
87	Rigorous
88	Average Rigor
89	Average Rigor
90	Easy
Easy 20%	1,4,6,7,8,10,15,16,26,37,59,64,69,76,83,84,86,90
Average Rigor 40%	2,3,5,9,12,13,14,18,19,20,21,22,23,24,25,28,29,31,38,43, 47,52,54,55,58,60,66,71,72,73,74,77,78,81,88,89
Rigorous 40%	11,17,27,30,32,33,34,35,36,39,40,41,42,44,45,46,48, 49,50,51,53,56,57,61,62,63,65,67,68,70,75,79,80,82,85,87

Rationales with Sample Questions

1. **When you begin by assuming the conclusion of a theorem is false, then show that through a sequence of logically correct steps you contradict an accepted fact, this is known as**
   (Easy)(Skill 0001)

   A) inductive reasoning
   B) direct proof
   C) indirect proof
   D) exhaustive proof

Answer: C
By definition this describes the procedure of an indirect proof.

2. **Given the series of examples below, what is 5¢4?**

   $4¢3=13$ $\qquad$ $7¢2=47$
   $3¢1=8$ $\qquad$ $1¢5=-4$

   (Average Rigor)(Skill 0001)

   A) 20
   B) 29
   C) 1
   D) 21

Answer: D
By observation of the examples given, $a \not\subset b = a^2 - b$. Therefore,
$5 \not\subset 4 = 25 - 4 = 21$.

3. **A group of students working with trigonometric identities have concluded that $\cos 2x = 2\cos x$. How could you best lead them to discover their error?**
   (Average Rigor)(Skill 0001)

   A) Have the students plug in values on their calculators.
   B) Direct the student to the appropriate chapter in the text.
   C) Derive the correct identity on the board.
   D) Provide each student with a table of trig identities.

Answer: C

The personal approach of answer C is the best way to help students discover their error. By demonstrating the correct process of derivation of the appropriate identity on the board, students will be able to learn both the correct answer and the correct method for arriving at the answer.

4. **Which best describes the type of error observed below?**
   (Easy)(Skill 0001)

-3 + 7 = -4	6(-10) = - 60
-5(-15) = 75	-3+-8 = 11
8-12 = -4	7- -8 = 15

   A) The student is incorrectly multiplying integers.
   B) The student has incorrectly applied rules for adding integers to subtracting integers.
   C) The student has incorrectly applied rules for multiplying integers to adding integers.
   D) The student is incorrectly subtracting integers.

Answer: C
The errors are in the following: -3+7=-4 and –3 + -8 = 11, where the student seems to be using the rules for signs when multiplying, instead of the rules for signs when adding.

5. **About two weeks after introducing formal proofs, several students in your geometry class are having a difficult time remembering the names of the postulates. They cannot complete the reason column of the proof and as a result are not even attempting the proofs. What would be the best approach to help students understand the nature of geometric proofs?**
   (Average Rigor)(Skill 0001)

   A) Give them more time; proofs require time and experience.
   B) Allow students to write an explanation of the theorem in the reason column instead of the name.
   C) Have the student copy each theorem in a notebook.
   D) Allow the students to have open book tests.

Answer: B
The best way to encourage the students and to help them master the subject is to allow them to give an explanation of the concept or substance of the theorem when they are unable to recall the name of the theorem. Although learning the names of theorems has some value, it is far more important that students master the substance of the theorems. If such mastery is demonstrated on tests, little or no penalty should be applied, as an understanding of geometric proofs has by and large been shown.

6. **Which mathematician is best known for his work in developing non-Euclidean geometry?**
   (Easy)(Skill 0002)

   A) Descartes
   B) Riemann
   C) Pascal
   D) Pythagoras

Answer: B
In the mid-nineteenth century, Reimann and other mathematicians developed elliptic geometry.

7. **According to Piaget, what stage in a student's development would be appropriate for introducing abstract concepts in geometry?**
   (Easy)(Skill 0002)

   A) concrete operational
   B) formal operational
   C) sensori-motor
   D) pre-operational

Answer: A

The concrete operational stage is characteristic of ages 7 - 11. In this stage children start to think more abstractly and can look at the general characteristics of mathematical objects.

The sensori-motor stage applies to children from the age of 18 months to 2 years. The pre-operational stage refers to children aged 2 - 7 and the formal operational stage is for children in upper elementary.

8. **Change $.\overline{63}$ into a fraction in simplest form.**
   (Easy)(Skill 0002)

   A) $63/100$
   B) $7/11$
   C) $6\ 3/10$
   D) $2/3$

Answer: B
Let N = .636363…. Then multiplying both sides of the equation by 100 or 10^2 (because there are 2 repeated numbers), we get 100N = 63.636363… Then subtracting the two equations gives 99N = 63 or N = $\dfrac{63}{99} = \dfrac{7}{11}$.

9. A student had 60 days to appeal the results of an exam. If the results were received on March 23, what was the last day that the student could appeal?
   (Average Rigor)(Skill 0002)

   A) May 21
   B) May 22
   C) May 23
   D) May 24

Answer: B
Recall: 30 days in April and 31 in March. 8 days in March + 30 days in April + 22 days in May brings him to a total of 60 days on May 22.

10. Express .0000456 in scientific notation.
    (Easy)(Skill 0002)

    A) $4.56x10^{-4}$
    B) $45.6x10^{-6}$
    C) $4.56x10^{-6}$
    D) $4.56x10^{-5}$

Answer: D
In scientific notation, the decimal point belongs to the right of the 4, the first significant digit. To get from 4.56×10^{-5} back to 0.0000456, we would move the decimal point 5 places to the left.

11. The volume of water flowing through a pipe varies directly with the square of the radius of the pipe. If the water flows at a rate of 80 liters per minute through a pipe with a radius of 4 cm, at what rate would water flow through a pipe with a radius of 3 cm?
   (Rigorous)(Skill 0003)

   A) 45 liters per minute
   B) 6.67 liters per minute
   C) 60 liters per minute
   D) 4.5 liters per minute

Answer: A

Set up the direct variation: $\dfrac{V}{r^2} = \dfrac{V}{r^2}$. Substituting gives $\dfrac{80}{16} = \dfrac{V}{9}$. Solving for V gives 45 liters per minute.

12. The mass of a Chips Ahoy cookie would be to
   (Average Rigor)(Skill 0003)

   A) 1 kilogram
   B) 1 gram
   C) 15 grams
   D) 15 milligrams

Answer: C

Since an ordinary cookie would not weigh as much as 1 kilogram, or as little as 1 gram or 15 milligrams, the only reasonable answer is 15 grams.

13. What would be the least appropriate use for handheld calculators in the classroom?
   (Average Rigor)(Skill 0003)

   A) practice for standardized tests
   B) integrating algebra and geometry with applications
   C) justifying statements in geometric proofs
   D) applying the law of sines to find dimensions

Answer: C

There is no need for calculators when justifying statements in a geometric proof.

14. What would be the total cost of a suit for $295.99 and a pair of shoes for $69.95 including 6.5% sales tax?
(Average Rigor)(Skill 0003)

   A) $389.73
   B) $398.37
   C) $237.86
   E) $315.23

Answer: A

Before the tax, the total comes to $365.94. Then .065(365.94) = 23.79. With the tax added on, the total bill is 365.94 + 23.79 = $389.73. (Quicker way: 1.065(365.94) = 389.73.)

15. Which of the following is the best example of the value of personal computers in advanced high school mathematics?
(Easy)(Skill 0003)

   A) Students can independently drill and practice test questions.
   B) Students can keep an organized list of theorems and postulates on a word processing program.
   C) Students can graph and calculate complex functions to explore their nature and make conjectures.
   D) Students are better prepared for business because of mathematics computer programs in high school.

Answer: C

16. Which of the following illustrates an inverse property?
(Easy)(Skill 0004)

   A) $a + b = a - b$
   B) $a + b = b + a$
   C) $a + 0 = a$
   D) $a + (-a) = 0$

Answer: D

Because $a + (-a) = 0$ is a statement of the Additive Inverse Property of Algebra.

17. Simplify: $\sqrt{27} + \sqrt{75}$
    (Rigorous)(Skill 0004)

    A) $8\sqrt{3}$
    B) 34
    C) $34\sqrt{3}$
    D) $15\sqrt{3}$

Answer: A
Simplifying radicals gives $\sqrt{27} + \sqrt{75} = 3\sqrt{3} + 5\sqrt{3} = 8\sqrt{3}$.

18. Determine the number of subsets of set *K*.
    $$K = \{4, 5, 6, 7\}$$
    (Average Rigor)(Skill 0004)

    A) 15
    B) 16
    C) 17
    D) 18

Answer: B
A set of n objects has 2^n subsets. Therefore, here we have $2^4 = 16$ subsets. These subsets include four which each have 1 element only, six which each have 2 elements, four which each have 3 elements, plus the original set, and the empty set.

19. Simplify: $\dfrac{10}{1+3i}$
    (Average Rigor)(Skill 0004)

    A) $-1.25(1-3i)$
    B) $1.25(1+3i)$
    C) $1+3i$
    D) $1-3i$

Answer: D
Multiplying numerator and denominator by the conjugate gives

$$\frac{10}{1+3i} \times \frac{1-3i}{1-3i} = \frac{10(1-3i)}{1-9i^2} = \frac{10(1-3i)}{1-9(-1)} = \frac{10(1-3i)}{10} = 1-3i.$$

20. Which of the following sets is closed under division?
   (Average Rigor)(Skill 0004)

   I) {½, 1, 2, 4}
   II) {-1, 1}
   III) {-1, 0, 1}

   A) I only
   B) II only
   C) III only
   D) I and II

Answer: B

I is not closed because $\dfrac{4}{.5} = 8$ and 8 is not in the set.

III is not closed because $\dfrac{1}{0}$ is undefined.

II is closed because $\dfrac{-1}{1} = -1, \dfrac{1}{-1} = -1, \dfrac{1}{1} = 1, \dfrac{-1}{-1} = 1$ and all the answers are in the set.

21. Identify the correct sequence of subskills required for solving and graphing inequalities involving absolute value in one variable, such as $|x+1| \le 6$.
   (Average Rigor)(Skill 0005)

   A) understanding absolute value, graphing inequalities, solving systems of equations
   B) graphing inequalities on a Cartesian plane, solving systems of equations, simplifying expressions with absolute value
   C) plotting points, graphing equations, graphing inequalities
   D) solving equations with absolute value, solving inequalities, graphing conjunctions and disjunctions

Answer: D
The steps listed in answer D would look like this for the given example:
If $|x + 1| \le 6$, then $-6 \le x + 1 \le 6$, which means $-7 \le x \le 5$. Then the inequality would be graphed on a numberline and would show that the solution set is all real numbers between −7 and 5, including −7 and 5.

22. Which of the following is always composite if x is odd, y is even, and both x and y are greater than or equal to 2?
(Average Rigor)(Skill 0005)

A) $x + y$

B) $3x + 2y$

C) $5xy$

D) $5x + 3y$

Answer: C
A composite number is a number which is not prime. The prime number sequence begins 2,3,5,7,11,13,17,.... To determine which of the expressions is always composite, experiment with different values of x and y, such as x=3 and y=2, or x=5 and y=2. It turns out that 5xy will always be an even number, and therefore, composite, if y=2.

23. Solve for v_0 : $d = at(v_t - v_0)$
(Average Rigor)(Skill 0005)

A) $v_0 = atd - v_t$

B) $v_0 = d - atv_t$

C) $v_0 = atv_t - d$

D) $v_0 = (atv_t - d)/at$

Answer: D
Using the Distributive Property and other properties of equality to isolate v_0 gives
$d = atv_t - atv_0, \quad atv_0 = atv_t - d, \quad v_0 = \dfrac{atv_t - d}{at}$.

24. $f(x) = 3x - 2; \ f^{-1}(x) =$
(Average Rigor)(Skill 0006)

A) $3x + 2$

B) $x/6$

C) $2x - 3$

D) $(x + 2)/3$

Answer: D
To find the inverse, $f^{-1}(x)$, of the given function, reverse the variables in the given equation, y = 3x – 2, to get x = 3y – 2. Then solve for y as follows:
x+2 = 3y, and y = $\dfrac{x+2}{3}$.

25. State the domain of the function $f(x) = \dfrac{3x-6}{x^2-25}$

    (Average Rigor)(Skill 0006)

    A) $x \neq 2$
    B) $x \neq 5, -5$
    C) $x \neq 2, -2$
    D) $x \neq 5$

Answer: B
The values of 5 and –5 must be omitted from the domain of all real numbers because if x took on either of those values, the denominator of the fraction would have a value of 0, and therefore the fraction would be undefined.

26. What is the equation of the graph below?
    (Easy)(Skill 0006)

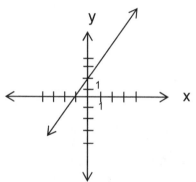

    A) $2x + y = 2$
    B) $2x - y = -2$
    C) $2x - y = 2$
    D) $2x + y = -2$

Answer: B
By observation, we see that the graph has a y-intercept of 2 and a slope of 2/1 = 2. Therefore its equation is y = mx + b = 2x + 2. Rearranging the terms gives 2x − y = -2.

27. **Which of the following is a factor of** $6 + 48m^3$
   (Rigorous)(Skill 0007)

   A) $(1 + 2m)$
   B) $(1 - 8m)$
   C) $(1 + m - 2m)$
   D) $(1 - m + 2m)$

Answer: A
Removing the common factor of 6 and then factoring the sum of two cubes gives
$6 + 48m^3 = 6(1 + 8m^3) = 6(1 + 2m)(1^2 - 2m + (2m)^2)$.

28. **Given** $f(x) = 3x - 2$ **and** $g(x) = x^2$, **determine** $g(f(x))$.
   (Average Rigor)(Skill 0007)

   A) $3x^2 - 2$
   B) $9x^2 + 4$
   C) $9x^2 - 12x + 4$
   D) $3x^3 - 2$

Answer: C
The composite function $g(f(x)) = (3x-2)^2 = 9x^2 - 12x + 4$.

29. Which graph represents the equation of $y = x^2 + 3x$ **?**
   (Average Rigor)(Skill 0007)

A)

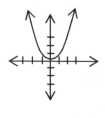

B)

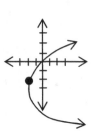

C)

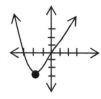

D)

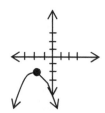

Answer: C
B is not the graph of a function. D is the graph of a parabola where the coefficient of x^2 is negative. A appears to be the graph of $y = x^2$. To find the x-intercepts of $y = x^2 + 3x$, set y = 0 and solve for x: $0 = x^2 + 3x = x(x + 3)$ to get x = 0 or x = -3. Therefore, the graph of the function intersects the x-axis at x=0 and x=-3.

30. Which of the following is incorrect?
   (Rigorous)(Skill 0007)

   A) $(x^2 y^3)^2 = x^4 y^6$
   B) $m^2(2n)^3 = 8m^2 n^3$
   C) $(m^3 n^4)/(m^2 n^2) = mn^2$
   D) $(x + y^2)^2 = x^2 + y^4$

Answer: D
Using FOIL to do the expansion, we get $(x + y^2)^2 = (x + y^2)(x + y^2) = x^2 + 2xy^2 + y^4$.

31. Which graph represents the solution set for $x^2 - 5x > -6$?
(Average Rigor)(Skill 0007)

A)
    -2 0 2

B)
    -3   0

C)
    -2 0 2

D)
    -3   0 2 3

Answer: D
Rewriting the inequality gives $x^2 - 5x + 6 > 0$. Factoring gives $(x - 2)(x - 3) > 0$. The two cut-off points on the number line are now at $x = 2$ and $x = 3$. Choosing a random number in each of the three parts of the numberline, we test them to see if they produce a true statement. If $x = 0$ or $x = 4$, $(x-2)(x-3)>0$ is true. If $x = 2.5$, $(x-2)(x-3)>0$ is false. Therefore the solution set is all numbers smaller than 2 or greater than 3.

32. What would be the seventh term of the expanded binomial $(2a+b)^8$?
(Rigorous)(Skill 0007)

A) $2ab^7$
B) $41a^4b^4$
C) $112a^2b^6$
D) $16ab^7$

Answer: C

The set-up for finding the seventh term is $\dfrac{8(7)(6)(5)(4)(3)}{6(5)(4)(3)(2)(1)}(2a)^{8-6}b^6$ which gives

$28(4a^2b^6)$ or $112a^2b^6$.

33. **Find the zeroes of** $f(x) = x^3 + x^2 - 14x - 24$

    (Rigorous)(Skill 0007)

    A) 4, 3, 2
    B) 3, -8
    C) 7, -2, -1
    D) 4, -3, -2

Answer: D
Possible rational roots of the equation $0 = x^3 + x^2 - 14x - 24$ are all the positive and negative factors of 24. By substituting into the equation, we find that -2 is a root, and therefore that x+2 is a factor. By performing the long division $(x^3 + x^2 - 14x - 24)/(x+2)$, we can find that another factor of the original equation is $x^2 - x - 12$ or $(x-4)(x+3)$. Therefore the zeros of the original function are -2, -3, and 4.

34. **Evaluate** $3^{1/2}(9^{1/3})$

    (Rigorous)(Skill 0008)

    A) $27^{5/6}$
    B) $9^{7/12}$
    C) $3^{5/6}$
    D) $3^{6/7}$

Answer: B
Getting the bases the same gives us $3^{\frac{1}{2}}3^{\frac{2}{3}}$. Adding exponents gives $3^{\frac{7}{6}}$. Then some additional manipulation of exponents produces $3^{\frac{7}{6}} = 3^{\frac{14}{12}} = \left(3^2\right)^{\frac{7}{12}} = 9^{\frac{7}{12}}$.

35. **Solve for** x: $18 = 4 + |2x|$

    (Rigorous)(Skill 0008)

    A) $\{-11, 7\}$
    B) $\{-7, 0, 7\}$
    C) $\{-7, 7\}$
    D) $\{-11, 11\}$

Answer: C
Using the definition of absolute value, two equations are possible: $18 = 4 + 2x$ or $18 = 4 - 2x$. Solving for x gives x = 7 or x = -7.

36. Which equation corresponds to the logarithmic statement:
$\log_x k = m$?
(Rigorous)(Skill 0009)
   A) $x^m = k$
   B) $k^m = x$
   C) $x^k = m$
   D) $m^x = k$

Answer: A
By definition of log form and exponential form, $\log_x k = m$ corresponds to $x^m = k$.

37. Which of the following is the best approximate value of x in the following equation?

$$\ln(x) = 8$$

(Easy)(Skill 0009)

   A) 2981
   B) -2981
   C) 2.079
   D) -2.079

Answer: A

This is a straightforward problem, requiring that we let both side of the equation be exponents of the base e:

$$e^{\ln(x)} = e^8$$

$$x = e^8 \approx 2981$$

38. Solve for x in the following equation:

$$a\log(3x) = b$$

(Average Rigor)(Skill 0009)

A) $x = 30^{\frac{b}{a}}$

B) $x = \dfrac{10^{\frac{a}{b}}}{3}$

C) $x = \left(\dfrac{10}{3}\right)^{\frac{b}{a}}$

D) $x = \dfrac{10^{\frac{b}{a}}}{3}$

Answer: D

The following manipulations are required:

$$\log(3x) = \frac{b}{a}$$

$$3x = 10^{\frac{b}{a}}$$

$$x = \frac{10^{\frac{b}{a}}}{3}$$

39. The populations of two towns are growing exponentially as functions of time:

$$\textbf{Town 1: } P(t) = 100e^{0.010t}$$

$$\textbf{Town 2: } P(t) = 120e^{0.008t}$$

Where t is positive and given in years. It t=0 in the year 2000, in what year will the towns have the same population?

(Rigorous)(Skill 0009)

   A) 2044
   B) 2091
   C) 2910
   D) 2050

Answer: B

We will call t' the year when the populations of the two towns are equal and set the two expressions equal:

$$100e^{0.010t'} = 120e^{0.008t'}$$

$$\frac{e^{0.010t'}}{e^{0.008t'}} = \frac{120}{100}$$

$$e^{0.010t'-0.008t'} = 1.2$$

$$\ln(e^{0.002t'}) = \ln(1.2)$$

$$0.002t' = \ln(1.2)$$

$$t' = \frac{\ln(1.2)}{0.002} \approx 91$$

Thus the populations will be equal after 91 years, or in 2091.

40. Which expression is equivalent to $1 - \sin^2 x$?
    (Rigorous)(Skill 0010)

    A) $1 - \cos^2 x$
    B) $1 + \cos^2 x$
    C) $1/\sec x$
    D) $1/\sec^2 x$

Answer: D
Using the Pythagorean Identity, we know $\sin^2 x + \cos^2 x = 1$. Thus $1 - \sin^2 x = \cos^2 x$, which by definition is equal to $1/\sec^2 x$.

41. Find the slope of the line tangent to $y = 3x(\cos x)$ at $(\pi/2, \pi/2)$.
    (Rigorous)(Skill 0010)

    A) $-3\pi/2$
    B) $3\pi/2$
    C) $\pi/2$
    D) $-\pi/2$

Answer: A
To find the slope of the tangent line, find the derivative, and then evaluate it at x $= \dfrac{\pi}{2}$. y' = 3x(-sinx)+3cosx. At the given value of x,

$$y' = 3(\frac{\pi}{2})(-\sin\frac{\pi}{2}) + 3\cos\frac{\pi}{2} = \frac{-3\pi}{2}.$$

42. Find the equation of the line tangent to $y = 3x^2 - 5x$ at $(1, -2)$.
    (Rigorous)(Skill 0010)

    A) $y = x - 3$
    B) $y = 1$
    C) $y = x + 2$
    D) $y = x$

Answer: A
To find the slope of the tangent line, find the derivative, and then evaluate it at x=1.

y'=6x-5=6(1)-5=1. Then using point-slope form of the equation of a line, we get y+2=1(x-1) or y = x-3.

43. Determine the measures of angles A and B.
(Average Rigor)(Skill 0011)

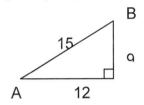

A) A = 30°, B = 60°
B) A = 60°, B = 30°
C) A = 53°, B = 37°
D) A = 37°, B = 53°

Answer: D
Tan A = 9/12=.75 and tan^{-1}.75 = 37 degrees. Since angle B is complementary to angle A, the measure of angle B is therefore 53 degrees.

44. What is the measure of minor arc AD, given measure of arc PS is 40°
and $m < K = 10°$?
(Rigorous)(Skill 0011)

A) 50°
B) 20°
C) 30°
D) 25°

Answer: B
The formula relating the measure of angle K and the two arcs it intercepts is
$m\angle K = \dfrac{1}{2}(mPS - mAD)$. Substituting the known values, we get $10 = \dfrac{1}{2}(40 - mAD)$.
Solving for mAD gives an answer of 20 degrees.

45. Find the first derivative of the function: $f(x) = x^3 - 6x^2 + 5x + 4$
(Rigorous)(Skill 0012)

A) $3x^3 - 12x^2 + 5x = f'(x)$
B) $3x^2 - 12x - 5 = f'(x)$
C) $3x^2 - 12x + 9 = f'(x)$
D) $3x^2 - 12x + 5 = f'(x)$

Answer: D
Use the Power Rule for polynomial differentiation: if y = ax^n, then y'=nax^{n-1}.

46. Differentiate: $y = e^{3x+2}$
   (Rigorous)(Skill 0012)

   A) $3e^{3x+2} = y'$

   B) $3e^{3x} = y'$

   C) $6e^3 = y'$

   D) $(3x+2)e^{3x+1} = y'$

Answer: A
Use the Exponential Rule for derivatives of functions of e: if y = ae$^{f(x)}$, then
y' = f'(x)ae$^{f(x)}$. **Answer is A. (Rigorous)**

47. How does the function $y = x^3 + x^2 + 4$ **behave from** $x = 1$ **to** $x = 3$?
   (Average Rigor)(Skill 0012)

   A) increasing, then decreasing
   B) increasing
   C) decreasing
   D) neither increasing nor decreasing

Answer: B
To find critical points, take the derivative, set it equal to 0, and solve for x.
f'(x) = 3x^2 + 2x = x(3x+2)=0. CP at x=0 and x=-2/3. Neither of these CP is on the
interval from x=1 to x=3. Testing the endpoints: at x=1, y=6 and at x=3, y=38.
Since the derivative is positive for all values of x from x=1 to x=3, the curve is
increasing on the entire interval.

48. Find the absolute maximum obtained by the function $y = 2x^2 + 3x$ **on the**
   interval $x = 0$ **to** $x = 3$.
   (Rigorous)(Skill 0012)

   A) $-3/4$
   B) $-4/3$
   C) 0
   D) 27

Answer: D
Find CP at x=-.75 as shown above. Since the CP is not in the interval from x=0
to x=3, just find the values of the functions at the endpoints. When x=0, y=0, and
when x=3, y = 27. Therefore 27 is the absolute maximum on the given interval.

49. Find the antiderivative for $4x^3 - 2x + 6 = y$.
(Rigorous)(Skill 0012)

A) $x^4 - x^2 + 6x + C$
B) $x^4 - 2/3x^3 + 6x + C$
C) $12x^2 - 2 + C$
D) $4/3x^4 - x^2 + 6x + C$

Answer: A

Use the rule for polynomial integration: given ax^n, the antiderivative is $\dfrac{ax^{n+1}}{n+1}$.

50. Evaluate $\int_0^2 (x^2 + x - 1)dx$
(Rigorous)(Skill 0012)

A) 11/3
B) 8/3
C) -8/3
D) -11/3

Answer: B

Use the fundamental theorem of calculus to find the definite integral: given a continuous function f on an interval [a,b], then $\int_a^b f(x)dx = F(b) - F(a)$, where F is an antiderivative of f.

$\int_0^2 (x^2 + x - 1)dx = (\dfrac{x^3}{3} + \dfrac{x^2}{2} - x)$ Evaluate the expression at x=2, at x=0, and then subtract to get 8/3 + 4/2 – 2-0 = 8/3.

51. Evaluate: $\int (x^3 + 4x - 5)dx$
(Rigorous)(Skill 0012)

A) $3x^2 + 4 + C$
B) $\dfrac{1}{4}x^4 - 2/3x^3 + 6x + C$
C) $x^{4/3} + 4x - 5x + C$
D) $x^3 + 4x^2 - 5x + C$

Answer: B

Integrate as in the preceding problem.

52. Find the area under the function $y = x^2 + 4$ from $x = 3$ to $x = 6$.
    (Average Rigor)(Skill 0012)

    A) 75
    B) 21
    C) 96
    D) 57

Answer: A

To find the area set up the definite integral: $\int_{3}^{6}(x^2 + 4)dx = (\frac{x^3}{3} + 4x)$. Evaluate the expression at x=6, at x=3, and then subtract to get (72+24)-(9+12)=75.

53. Find the antiderivative for the function $y = e^{3x}$.
    (Rigorous)(Skill 0012)

    A) $3x(e^{3x}) + C$
    B) $3(e^{3x}) + C$
    C) $1/3(e^x) + C$
    D) $1/3(e^{3x}) + C$

Answer: D
Use the rule for integration of functions of e: $\int e^x dx = e^x + C$.

54. The acceleration of a particle is dv/dt = 6 m/s². Find the velocity at t=10 given an initial velocity of 15 m/s.
    (Average Rigor)(Skill 0013)

    A) 60 m/s
    B) 150 m/s
    C) 75 m/s
    D) 90 m/s

Answer: C
Recall that the derivative of the velocity function is the acceleration function. In reverse, the integral of the acceleration function is the velocity function. Therefore, if a=6, then v=6t+C. Given that at t=0, v=15, we get v = 6t+15. At t=10, v=60+15=75m/s.

55. If the velocity of a body is given by v = 16 - t², find the distance traveled from t = 0 until the body comes to a complete stop.
(Average Rigor)(Skill 0013)

A) 16
B) 43
C) 48
D) 64

Answer: B
Recall that the derivative of the distance function is the velocity function. In reverse, the integral of the velocity function is the distance function. To find the time needed for the body to come to a stop when v=0, solve for t: v = 16 − t² = 0.

Result: t = 4 seconds. The distance function is s = 16t - $\dfrac{t^3}{3}$. At t=4, s= 64 – 64/3

or approximately 43 units.

56. Compute the area of the shaded region, given a radius of 5 meters. 0 is the center.
(Rigorous)(Skill 0014)

A) 7.13 cm²
B) 7.13 m²
C) 78.5 m²
D) 19.63 m²

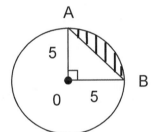

Answer: B

Area of triangle AOB is .5(5)(5) = 12.5 square meters. Since $\dfrac{90}{360} = .25$, the area

of sector AOB (pie-shaped piece) is approximately .25(π)5² = 19.63. Subtracting the triangle area from the sector area to get the area of segment AB, we get approximately 19.63-12.5 = 7.13 square meters.

57. If the area of the base of a cone is tripled, the volume will be
    (Rigorous)(Skill 0014)

    A) the same as the original
    B) 9 times the original
    C) 3 times the original
    D) 3 π times the original

Answer: C

The formula for the volume of a cone is $V = \frac{1}{3}Bh$, where B is the area of the

circular base and h is the height. If the area of the base is tripled, the volume
becomes

$V = \frac{1}{3}(3B)h = Bh$, or three times the original area.

58. Choose the correct statement concerning the median and altitude in a
    triangle.
    (Average Rigor)(Skill 0014)

    A) The median and altitude of a triangle may be the same segment.
    B) The median and altitude of a triangle are always different segments.
    C) The median and altitude of a right triangle are always the same
       segment.
    D) The median and altitude of an isosceles triangle are always the same
       segment.

Answer: A

The most one can say with certainty is that the median (segment drawn to the
midpoint of the opposite side) and the altitude (segment drawn perpendicular to
the opposite side) of a triangle <u>may</u> coincide, but they more often do not. In an
isosceles triangle, the median and the altitude to the <u>base</u> are the same
segment.

59. Find the surface area of a box which is 3 feet wide, 5 feet tall, and 4 feet deep.
(Easy)(Skill 0014)

   A) 47 sq. ft.
   B) 60 sq. ft.
   C) 94 sq. ft
   D) 188 sq. ft.

Answer: C
Let's assume the base of the rectangular solid (box) is 3 by 4, and the height is 5. Then the surface area of the top and bottom together is 2(12) = 24. The sum of the areas of the front and back are 2(15) = 30, while the sum of the areas of the sides are 2(20)=40. The total surface area is therefore 94 square feet.

60. Given a 30 meter x 60 meter garden with a circular fountain with a 5 meter radius, calculate the area of the portion of the garden not occupied by the fountain.
(Average Rigor)(Skill 0014)

   A) 1721 m²
   B) 1879 m²
   C) 2585 m²
   D) 1015 m²

Answer: A
Find the area of the garden and then subtract the area of the fountain:
$30(60) - \pi(5)^2$ or approximately 1721 square meters.

61. Determine the area of the shaded region of the trapezoid in terms of x and y.
    (Rigorous)(Skill 0014)

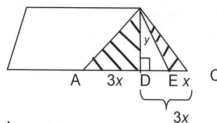

A) $4xy$

B) $2xy$

C) $3x^2y$

D) There is not enough information given.

Answer: B
To find the area of the shaded region, find the area of triangle ABC and then subtract the area of triangle DBE. The area of triangle ABC is .5(6x)(y) = 3xy. The area of triangle DBE is .5(2x)(y) = xy. The difference is 2xy.

62. Find the area of the figure pictured below.
    (Rigorous)(Skill 0014)

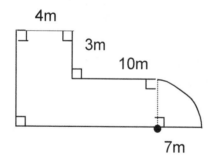

A) 136.47 m²

B) 148.48 m²

C) 293.86 m²

D) 178.47 m²

Answer: B
Divide the figure into 2 rectangles and one quarter circle. The tall rectangle on the left will have dimensions 10 by 4 and area 40. The rectangle in the center will have dimensions 7 by 10 and area 70. The quarter circle will have area .25(π)7² = 38.48.
The total area is therefore approximately 148.48.

63. Choose the diagram which illustrates the construction of a perpendicular to the line at a given point on the line. (Rigorous)(Skill 0015)

A)

B)

C)

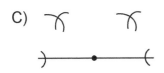

D)

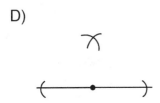

Answer: D
Given a point on a line, place the compass point there and draw two arcs intersecting the line in two points, one on either side of the given point. Then using any radius larger than half the new segment produced, and with the pointer at each end of the new segment, draw arcs which intersect above the line. Connect this new point with the given point.

64. Which term most accurately describes two coplanar lines without any common points? (Easy)(Skill 0015)

A) perpendicular
B) parallel
C) intersecting
D) skew

Answer: B
By definition, parallel lines are coplanar lines without any common points.

65. **What is the degree measure of an interior angle of a regular 10 sided polygon?**
    (Rigorous)(Skill 0015)

    A) 18°
    B) 36°
    C) 144°
    D) 54°

Answer: C
Formula for finding the measure of each interior angle of a regular polygon with n sides is $\frac{(n-2)180}{n}$. For n=10, we get $\frac{8(180)}{10} = 144$.

66. **Which theorem can be used to prove $\triangle BAK \cong \triangle MKA$?**
    (Average Rigor)(Skill 0015)

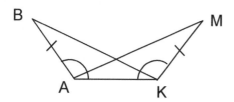

    A) SSS
    B) ASA
    C) SAS
    D) AAS

Answer: C
Since side AK is common to both triangles, the triangles can be proved congruent by using the Side-Angle-Side Postulate.

67. **Find the length of the major axis of $x^2 + 9y^2 = 36$.**
    (Rigorous)(Skill 0015)

    A) 4
    B) 6
    C) 12
    D) 8

Answer: C
Dividing by 36, we get $\frac{x^2}{36} + \frac{y^2}{4} = 1$, which tells us that the ellipse intersects the x-axis at 6 and –6, and therefore the length of the major axis is 12. (The ellipse intersects the y-axis at 2 and –2).

68. Which equation represents a circle with a diameter whose endpoints are $(0,7)$ and $(0,3)$?
   (Rigorous)(Skill 0016)

   A) $x^2 + y^2 + 21 = 0$
   B) $x^2 + y^2 - 10y + 21 = 0$
   C) $x^2 + y^2 - 10y + 9 = 0$
   D) $x^2 - y^2 - 10y + 9 = 0$

Answer: B
With a diameter going from (0,7) to (0,3), the diameter of the circle must be 4, the radius must be 2, and the center of the circle must be at (0,5). Using the standard form for the equation of a circle, we get $(x-0)^2 + (y-5)^2 = 2^2$. Expanding, we get $x^2 + y^2 - 10y + 21 = 0$.

69. Given that QO⊥NP and QO=NP, quadrilateral NOPQ can most accurately be described as a
   (Easy)(Skill 0016)

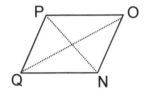

   A) parallelogram
   B) rectangle
   C) square
   D) rhombus

Answer: C
In an ordinary parallelogram, the diagonals are not perpendicular or equal in length. In a rectangle, the diagonals are not necessarily perpendicular. In a rhombus, the diagonals are not equal in length. In a square, the diagonals are both perpendicular and congruent.

70. Given $K(-4, y)$ and $M(2, -3)$ with midpoint $L(x, 1)$, determine the values of x and y.
    (Rigorous)(Skill 0016)

   A) $x = -1, \ y = 5$
   B) $x = 3, \ y = 2$
   C) $x = 5, \ y = -1$
   D) $x = -1, \ y = -1$

Answer: A
The formula for finding the midpoint (a,b) of a segment passing through the points $(x_1, y_1) \, and \, (x_2, y_2) \, is \, (a,b) = (\frac{x_1 + x_2}{2}, \frac{y_1 + y_2}{2})$. Setting up the corresponding equations from this information gives us $x = \frac{-4 + 2}{2}, and \, 1 = \frac{y - 3}{2}$. Solving for x and y gives x = -1 and y = 5.

71. Determine the rectangular coordinates of the point with polar coordinates (5, 60°).
    (Average Rigor)(Skill 0016)

   B) (0.5, 0.87)
   B) (-0.5, 0.87)
   C) (2.5, 4.33)
   D) (25, 150°)

Answer: C
Given the polar point (r, θ) = (5, 60), we can find the rectangular coordinates this way: (x,y) = $(r \cos \theta, r \sin \theta) = (5 \cos 60, 5 \sin 60) = (2.5, 4.33)$.

72. Given a vector with horizontal component 5 and vertical component 6, determine the length of the vector.
    (Average Rigor)(Skill 0017)

   A) 61
   B) $\sqrt{61}$
   C) 30
   D) $\sqrt{30}$

Answer: B
Using the Pythagorean Theorem, we get v = $\sqrt{36 + 25} = \sqrt{61}$.

73. Compute the distance from
    (-2,7) to the line $x = 5$.
    (Average Rigor)(Skill 0017)

    A) -9
    B) -7
    C) 5
    D) 7

Answer: D
The line $x = 5$ is a vertical line passing through (5,0) on the Cartesian plane. By observation the distance along the horizontal line from the point (-2,7) to the line x=5 is 7 units.

74. If a ship sails due south 6 miles, then due west 8 miles, how far was it from the starting point?
    (Average Rigor)(Skill 0017)

    A) 100 miles
    B) 10 miles
    C) 14 miles
    E) 48 miles

Answer: B
Draw a right triangle with legs of 6 and 8. Find the hypotenuse using the Pythagorean Theorem. $6^2 + 8^2 = c^2$. Therefore, c = 10 miles.

75. Find the sum of the first one hundred terms in the progression.

    (-6, -2, 2 . . .)
    (Rigorous)(Skill 0018)

    A) 19,200
    B) 19,400
    C) -604
    D) 604

Answer: A
To find the 100th term: $t_{100} = -6 + 99(4) = 390$. To find the sum of the first 100 terms: $S = \dfrac{100}{2}(-6 + 390) = 19200$.

76. What is the value of the following expression?

$$\frac{8!}{3!}$$

(Easy)(Skill 0018)

A) 13440
B) 120
C) 6720
D) 4.01

Answer: C

$$\frac{8!}{3!} = \frac{1 \times 2 \times 3 \times 4 \times 5 \times 6 \times 7 \times 8}{1 \times 2 \times 3} = 4 \times 5 \times 6 \times 7 \times 8 = 6720$$

77. How many ways are there to choose a potato and two green vegetables from a choice of three potatoes and seven green vegetables?
(Average Rigor)(Skill 00018)

A) 126
B) 63
C) 21
D) 252

Answer: A
There are 3 slots to fill. There are 3 choices for the first, 7 for the second, and 6 for the third. Therefore, the total number of choices is 3(7)(6) = 126.

78. A jar contains 6 red beans, 5 black beans, and 8 garbanzo beans. If a single bean is chosen at random, what the probability of it being a red bean?
(Average Rigor)(Skill 0019)

A) 18.6%
B) 25.6%
C) 52.3%
D) 31.5%

Answer: D

To determine the probability, we simply calculate the number of ways to choose red over divided by the number of way to choose any bean:

$$P = \frac{6}{6+5+8} = \frac{6}{19} = 0.315$$

79. A jar contains 6 red beans, 5 black beans, and 8 garbanzo beans. If two beans are drawn in sequence, what is the probability that first a red bean and then a black bean will be drawn (assume the red bean is not replaced after being drawn)?
(Rigorous)(Skill 0019)

A) 88.2%
B) 27.7%
C) 8.7%
D) 8.2%

Answer: C

To determine this, we simply multiply the probability that a red bean will be drawn by the probability that a black bean will be drawn (remembering to account for the loss of one red bean):

$$P = \left(\frac{6}{19}\right) \times \left(\frac{5}{18}\right) = 0.087$$

80. A jar contains 6 red beans, 5 black beans, and 8 garbanzo beans. Given that a red bean has been drawn (and not replaced), what is the probability that the will be black?
(Rigorous)(Skill 0019)

    A) 31.5%
    B) 8.7%
    C) 25.6%
    D) 27.7%

Answer: D

In this problem, we are to assume that one red bean has already been drawn and thus we must use conditional probability (Bayesian probability). Therefore the probability of the described event is the probability of drawing red then black divided by the probability of drawing red:

$$P = \frac{\left(\frac{6}{19}\right) \times \left(\frac{5}{18}\right)}{\frac{6}{19}} = 0.276$$

Note that these two events are independent, so this probability is the same as simply drawing a black bean $\left(\frac{5}{18}\right)$

81. The probability distribution shown below exhibits:
(Average Rigor)(Skill 0019)

    A) positive skew
    B) negative skew
    C) excessive kurtosis
    D) diminished kurtosis

Answer: A

Positive skew denotes that distribution has an elongated **right** tail (compared to what is expected in a normal distribution).

82. The number of pizza slices eaten per college student per year fits a normal distribution with a mean of 55 and a standard deviation of 15. The number of pizza slices eaten annually by the students in the top 2.5% of the distribution is greater than:
(Rigorous)(Skill 0019)

A) 70
B) 85
C) 100
D) 110

Answer: B

Recall that in a normal distribution, **95%** of the observations fall within 2 standard deviations of the mean. That is, they fall between:

$$\mu - 2\sigma \quad and \quad \mu + 2\sigma$$

Thus the top 2.5% will have a value greater than

$$\mu + 2\sigma$$

We can simply plug in the values provided in this problem to find:

$$\mu + 2\sigma = 55 + 2 \times 15 = 85$$

83. Compute the median for the following data set:

{12, 19, 13, 16, 17, 14}

(Easy)(Skill 0020)

A) 14.5
B) 15.17
C) 15
D) 16

Answer: C
Arrange the data in ascending order: 12,13,14,16,17,19. The median is the middle value in a list with an odd number of entries. When there is an even number of entries, the median is the mean of the two center entries. Here the average of 14 and 16 is 15.

84. Compute the standard deviation for the following set of temperatures.

(37, 38, 35, 37, 38, 40, 36, 39)

(Easy)(Skill 0020)

    A) 37.5
    B) 1.5
    C) 0.5
    D) 2.5

Answer: B
Find the mean: 300/8 = 37.5. Then, using the formula for standard deviation, we get

$$\sqrt{\dfrac{2(37.5-37)^2 + 2(37.5-38)^2 + (37.5-35)^2 + (37.5-40)^2 + (37.5-36)^2 + (37.5-39)^2}{8}}$$

which has a value of 1.5.

85. Half the students in a class scored 80% on an exam, most of the rest scored 85% except for one student who scored 10%. Which would be the best measure of central tendency for the test scores? (Rigorous)(Skill 0020)

    A) mean
    B) median
    C) mode
    D) either the median or the   mode because they are equal

Answer: B
In this set of data, the median (see #14) would be the most representative measure of central tendency since the median is independent of extreme values. Because of the 10% outlier, the mean (average) would be disproportionately skewed. In this data set, it is true that the median and the mode (number which occurs most often) are the same, but the median remains the best choice because of its special properties.

86. What conclusion can be drawn from the graph below?

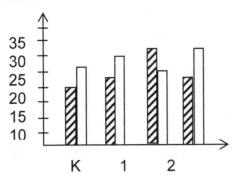

MLK Elementary
Student Enrollment   **Girls Boys**
(Easy)(Skill 0021)

A)   The number of students in first grade exceeds the number in second grade.
B)   There are more boys than girls in the entire school.
C)   There are more girls than boys in the first grade.
D)   Third grade has the largest number of students.

Answer: B
In Kindergarten, first grade, and third grade, there are more boys than girls. The number of extra girls in grade two is more than made up for by the extra boys in all the other grades put together.

87. Solve the system of equations for x, y **and** z.

$$3x + 2y - z = 0$$
$$2x + 5y = 8z$$
$$x + 3y + 2z = 7$$

(Rigorous)(Skill 0021)

A) $(-1,\ 2,\ 1)$
B) $(1,\ 2,\ -1)$
C) $(-3,\ 4,\ -1)$
D) $(0,\ 1,\ 2)$

Answer: A
Multiplying equation 1 by 2, and equation 2 by –3, and then adding together the two resulting equations gives -11y + 22z = 0. Solving for y gives y = 2z. In the meantime, multiplying equation 3 by –2 and adding it to equation 2 gives –y – 12z = -14. Then substituting 2z for y, yields the result z = 1. Subsequently, one can easily find that y = 2, and x = -1.

88. Evaluate the dot product of the following matrices as shown:

$$\begin{vmatrix} 1 & 5 \\ 7 & 3 \end{vmatrix} \bullet \begin{vmatrix} 4 & 5 \\ 2 & 9 \end{vmatrix}$$

(Average Rigor)(Skill 0021)

A) 70

B)
$$\begin{vmatrix} 4 & 50 \\ 34 & 67 \end{vmatrix}$$

C) 191

D)
$$\begin{vmatrix} 4 & 50 \\ 34 & 67 \end{vmatrix}$$

Answer:A

The dot product results in a scalar calculated as shown:

$$\begin{vmatrix} 1 & 5 \\ 7 & 3 \end{vmatrix} \bullet \begin{vmatrix} 4 & 5 \\ 2 & 9 \end{vmatrix} = (1*4) + (5*5) + (7*2) + (3*9) = 70$$

89. Find the value of the determinant of the matrix.
    (Average Rigor)(Skill 0021)

$$\begin{vmatrix} 2 & 1 & -1 \\ 4 & -1 & 4 \\ 0 & -3 & 2 \end{vmatrix}$$

A) 0
B) 23
C) 24
D) 40

Answer: C
To find the determinant of a matrix without the use of a graphing calculator, repeat the first two columns as shown,

2	1	-1	2	1
4	-1	4	4	-1
0	-3	2	0	-3

Starting with the top left-most entry, 2, multiply the three numbers in the diagonal going down to the right: 2(-1)(2)=-4. Do the same starting with 1: 1(4)(0)=0. And starting with −1: -1(4)(-3) = 12. Adding these three numbers, we get 8. Repeat the same process starting with the top right-most entry, 1. That is, multiply the three numbers in the diagonal going down to the left: 1(4)(2) = 8. Do the same starting with 2: 2(4)(-3) = -24 and starting with −1: -1(-1)(0) = 0. Add these together to get -16. To find the determinant, subtract the second result from the first: 8-(-16)=24.

90. What would be the shortest method of solution for the system of equations below?

$$3x + 2y = 38$$
$$4x + 8 = y$$

(Easy)(Skill 0021)

A) linear combination
B) additive inverse
C) substitution
D) graphing

Answer: C
Since the second equation is already solved for y, it would be easiest to use the substitution method.

MATHEMATICS 09   313

Constructed Response Answer

Problem:

The town of Verdant Slopes has been experiencing a boom in population growth. By the year 2000, the population had grown to 45,000, and by 2005, the population had reached 60,000.

a. Using the formula for slope as a model, find the average rate of change in population growth, expressing your answer in people per year.

b. Using the average rate of change determined in a., predict the population of Verdant Slopes in the year 2010.

Solution:

a. *Let t represent the time and p represent population growth. The two observances are represented by (t_1, p_1) and (t_2, p_2).*

1st observance = (t_1, p_1) = (2000, 45000)
2nd observance = (t_2, p_2) = (2005, 60000)

Use the formula for slope to find the average rate of change.

$$\text{Rate of change} = \frac{p_2 - p_1}{t_2 - t_1}$$

Substitute values.

$$= \frac{60000 - 45000}{2005 - 2000}$$

Simplify.

$$= \frac{15000}{5} = 3000 \, people \, / \, year$$

The average rate of change in population growth for Verdant Slopes between the years 2000 and 2005 was 3000 people/year.

b.

$$3000 \, people \, / \, year \times 5 \, years = 15000 \, people$$
$$60000 \, people + 15000 \, people = 75000 \, people$$

At a continuing average rate of growth of 3000 people/year, the population of Verdant Slopes could be expected to reach 75,000 by the year 2010.